SpringerBriefs in Electrical and Computer Engineering

For further volumes:
http://www.springer.com/series/10059

Jiuyong Li • Lin Liu • Thuc Duy Le

Practical Approaches to Causal Relationship Exploration

Springer

Jiuyong Li
School of Information Technology
  and Mathematical Sciences
University of South Australia
Adelaide, South Australia
Australia

Thuc Duy Le
School of Information Technology
  and Mathematical Sciences
University of South Australia
Adelaide, South Australia
Australia

Lin Liu
School of Information Technology
  and Mathematical Sciences
University of South Australia
Adelaide, South Australia
Australia

ISSN 2191-8112   ISSN 2191-8120 (electronic)
SpringerBriefs in Electrical and Computer Engineering
ISBN 978-3-319-14432-0   ISBN 978-3-319-14433-7 (eBook)
DOI 10.1007/978-3-319-14433-7

Library of Congress Control Number: 2015930835

Springer Cham Heidelberg New York Dordrecht London
© The Author(s) 2015

Springer is part of Springer Science+Business Media (www.springer.com)

Recommended by Xuemin (Sherman) Shen.

Preface

Causal discovery aims to discover the cause-effect relationships between variables. The relationships provide explanations as to how events have happened and predictions as to which events will happen in the future. Causality has been studied and utilised in almost all disciplines, e.g. medicine, epidemiology, biology, economics, physics, social science, as a basis for explanation, prediction and decision making.

Randomised controlled trials are the gold standard for discovering causal relationships. However, in many cases it is impossible to conduct randomised controlled trials due to cost, feasibility and/or ethical concerns.

With the rapid explosion of data collected in various areas, it is desirable to discover causal relationships in observational data. Causal discovery in data does not only reduce the costs for many scientific explorations and assist decision making, but importantly, it also helps detect crucial signals in data which might not be identified by domain experts to prevent serious consequences. Furthermore, data provides great opportunities for automated causal discovery and exploration by exploiting existing and developing new computational methods.

Significant achievements in causal modelling and inference have been made in various disciplines. Observational studies have long been used in medical research for identifying causal factors of diseases. Causal models, such as structural equation model and potential outcome model, have been used in a range of areas. In computer science, causal discovery based on graphical models has made significant theoretical achievements in the last 30 years.

However, there is still a lack of practical methods for causal discovery in large data sets. Most existing approaches are either hypothesis driven or unable to deal with large data sets. The causal discovery in this book refers to automated exploration of causal factors in large data sets without domain experts' hypotheses (domain experts may not know what to expect). More efficient methods are needed to deal with different types of data and applications for this purpose.

We aim to introduce four practical causal discovery methods for practitioners to mine their increasing data collection for causal information. We explain the mechanisms of the methods and provide demonstrations for the use of the methods. Relevant software tools can be found at their authors' home pages. Note that causal conclusions are not guaranteed by using a method since a data set may not satisfy the assumptions of the method, which itself includes heuristics. However, the methods in this book have a major advantage over other data mining or machine learning approaches for relationship exploration. These methods detect the relationship between two variables by considering other variables, and this consideration makes the discovered relationships less likely to be spurious or volatile.

We also aim to share our understandings in causal discovery with our peer researchers. Causal discovery is the goal for scientific exploration, and causal discovery in data is what computational researchers are able to contribute greatly to our society. Although it is arguable whether causality is definable or discoverable in data, the research endeavor in various disciplines has made significant progress in the theory of causal modelling and inference, which has shown promising applications in classification and prediction. It is time for data mining and machine learning researchers to reap the theoretical results and design efficient and practical algorithms for large scale causal relationship exploration. In this book, we characterise the causal discovery as a search problem—searching for persistent associations. This characterisation hopefully paves a short path for data mining and machine learning researchers to design more efficient algorithms for causal discovery with big data.

This book is also useful for students who want to improve their knowledge in this emerging area of research and applications. Causal discovery is a truly multiple disciplinary topic. We do not assume the knowledge of readers in a specific area. We try to explain the concepts and techniques clearly and use as many examples as possible. We hope that all readers, regardless of their knowledge backgrounds, can get some benefits by reading this book.

We thank Springer editors, Melissa Fearon and Jennifer Malat, for appreciating our idea and supporting us in writing the book. We also thank the Australian Research Council for its support for our work in this important area. This book would not be possible without the support of our collaborators, and here we just name a few: Jixue Liu, Zhou Jin, Bing-yu Sun and Rujing Wang. Algorithm implementation and experiments have been helped by Xin Zhu, Saisai Ma and Shu Hu. Certainly, the unfailing support of our families is crucial for us, and our gratitude is beyond words.

Adelaide, Australia,   *Jiuyong Li*
October 2014   *Lin Liu*
                                                                                               Thuc Duy Le

Contents

Chapter 1
Introduction

Abstract For centuries causal discovery has been an essential task for many disciplines due to the insights provided by causal relationships. However it is difficult and expensive to identify causal relationships with experimental approaches, especially when there are a large number of variables under consideration. Passively observed data thus has become an important source to be searched for causal relationships. The challenge of causal discovery with observational data lies in the fact that statistical associations detected from observational data are not necessarily causal. This book presents a number of computational methods for automated discovery of causal relationships around a given target (response) variable from large observational data sets, including high dimensional data sets. This chapter introduces the problem and the challenges, and outlines the ideas of the methods.

1.1 Finding Causal Relationships is Useful but Difficult

Although there is not a precise definition of causal relationships, the studies of causality are primarily on identifying the mechanisms of variables taking their values (e.g. why or what is the cause for a person to have a certain disease?), and how changes of cause variables lead to changes of outcome variables (e.g. how does the dose of a drug affect the treatment efficacy?). Therefore knowing events being causally related can be more useful than simply finding them being correlated or associated, because the knowledge of causes not only allows better explanations for events that have happened, but also enables the predication of the consequences of the changes of causes, thus to assist decision or policy making and possibly to prevent undesired events from occurring.

Randomised controlled trials (RCTs) [6] have long been recognised as the most effective approach to uncovering and validating causal relationships [18, 19]. However, it is often impossible to conduct such experiments due to ethical concerns or cost issues. For example, to study the potential causes of heart diseases (e.g. smoking or drinking), it will be unethical to require an experiment participant to smoke

© The Author(s) 2015   1
J. Li et al., *Practical Approaches to Causal Relationship Exploration*,
SpringerBriefs in Electrical and Computer Engineering, DOI 10.1007/978-3-319-14433-7_1

or drink. In some cases, e.g. in a life-threatening situation, RCTs will be totally prohibited.

Furthermore experimental approaches become very difficult and expensive to apply as the number of causes increases. For instance, gene knock-down experiments is an effective method to validate if one gene regulates the expression of another gene. However, given the huge number of genes (tens of thousands), it is extremely difficult to do the experiment for each pair of genes.

As a result, causal discovery with passively observed data (i.e. observational data) is often used when controlled experiments are not feasible, and this alternative has become increasingly attractive because of the abundance of data available in various fields.

However, finding causal relationships from observational data is a challenging task because passively observed data essentially does not capture the change of a variable as a result of the forced change of another variable. What we can obtain from such data are statistical associations, which may not be causal.

1.2 Associations and Causation

Statistical association analysis, especially correlation analysis with the well known correlation coefficients, such as Pearson correlation coefficient, has been a major technique for identifying useful relationships from observational data. With the widespread applications of association analysis over the years, a notion has also been formed that associations are not necessarily causal, as illustrated by the following examples.

- Associations may be accidental.

    Two completely irrelevant variables may be found statistically associated.

    For example, strong correlations were found between the number of storks and the number of the newborns [12, 14]. Although many of us were told in our childhood that we had been brought home to our parents by a white stork [17], obviously stork population and human births are irrelevant from a scientific point of view. There is even less chance for them to be causally related as it is impossible that the number of babies delivered would have changed if the stork population was manipulated.
- Associations may be spurious.

    Sometimes a correlation may seem to be genuine and indicative of a causal relationship, but it is indeed a result of some common cause of the associated variables.

    A frequently used example is the correlation between children's shoe sizes and their reading skills. From the data of some school children we may easily see a positive correlation between the children's shoe sizes and their reading skills. Such correlations may be observed repeatedly from the data collected from different schools. However, this does not mean that bigger shoe sizes (or foot size if children always get right-sized shoes from their parents) is a cause of better reading skills or vice versa since there is a common cause for children's shoe sizes and reading skills, their *age*. Generally children's shoe sizes and reading skills will both increase when they grow up.

- Associations may be conditional.

  There are situations when the associations observed in sub-populations or under certain conditions disappear or have their directions reversed in the aggregation. In such cases we cannot draw conclusions of causal relationships simply from these associations either.

  A famous illustration of association reversal is the study done by Bickel et al. on the 1973 admission data of University of Berkeley [4]. It was found that in the disaggregated data for individual departments (sub-populations), most departments favoured female applicants, i.e. more female applicants were admitted than male applicants in a department. However, in the aggregated data, an overall population of 12,763 applicants with 8442 males and 4321 females, male applicants are more likely to be admitted (44.3 %) than female applicants (34.5 %). There is a positive association between Gender and Admission Status in a sub-population, but a negative association in the overall population. In this case, we cannot derive a causal relationship between Gender and Admission Status. In fact, the explanation by Bickel et al. was that more female students had applied for departments that were more difficult to get in (i.e. with lower acceptance rate), resulting in a lower overall percentages of admission for them.

The above examples of spurious and conditional associations are phenomena of the much discussed Simpson's Paradox [3, 11, 16], which is often used to alert us to be cautious in using statistical associations for inferring causal relationships [3].

With all these considerations on associations, however, it is well accepted that causation implies associations, thus an association is a necessary condition for establishing a causal relationship. In the Hill's criteria for causality [5], a commonly used checklist by epidemiologists and medical researchers when assessing if an association should be considered as causation, strength of association is on the top of the list.

Therefore, it is reasonable to start causal discovery with association analysis. The question is, how to detect genuine causal relationships from associations.

1.3 The Practical Approaches to Causal Discovery

The causal discovery methods presented in this book justify the causal nature of an association between two variables on the basis of its *persistence*, that is, the association expected to exist in all situations without being affected by the values of other variables. For example, if it is observed that female workers, in different areas, with different qualifications and of different ages, always earn less than male workers, i.e. the association between being female workers and receiving low salaries is persistent, then it is reasonable to accept the causal relationship between being a female and receiving low salary. This view of causal relationships aligns with the second Hill's criterion for causality [5], which states that to establish an association as causation, it needs to be "repeatedly observed by different persons, in different places, circumstances and times".

Among the four methods introduced in this book, the first two, PC-simple [8] and HITON-PC [1, 2] are variants of the well-known PC algorithm for learning the structure of a causal Bayesian network [20]. The PC algorithm is named after its developer, Peter Spirts and Clark Glymour. The PC algorithm determines if the association between two variables is deemed to be causal using conditional independence tests, conditioning on the remaining variables. A pair of variables are considered to have no direct cause-effect relationship once a subset of the remaining variables are found such that conditioning on this subset of variables, the two variables are independent.

The next method, CR-PA [7] derives a causal relationship or a causal rule (CR) from an association through statistical partial association (PA) test, to check if the association is valid given any value assignment of all other variables. The CR-CS [9] method is also aimed at discovering causal rules from observational data. It exploits the idea of a cohort study (CS) and assesses if an association exists regardless of the values of other variables by using samples that have *matching* values for other variables.

The four methods are viewed as practical approaches to causal discovery due to the following considerations.

Firstly, all the methods have been developed for facilitating *automated* causal discoveries from potentially large and high dimensional data sets, and they all assume limited prior knowledge of the causes. In practice, large data sets with a huge number of records and/or variables are ubiquitous. With these data sets, we often have little idea about the potential causes of an effect (or target) of interest. In this case, given the large number of variables, it will be infeasible to use experimental approaches to detect the causes. Instead, it would be very useful to firstly apply some automated methods to generate high quality causal hypotheses around the target (response) variable using the data. The discovered causal relationships can then be used as the basis for further study, e.g. to guide the design of controlled experiments to validate the hypotheses. For example, given a gene expression data set with tens of thousands of variables (genes), we can use one of the methods to detect possible causal (gene regulatory) relationships between a certain gene (the target) and all other genes. Then the findings can be used to assist the design of gene knock-down experiments such that only those highly confident relationships are to be validated using experiments. Given this purpose, efficiency was a major concern when these causal discovery methods were developed, and the requirements on false discoveries are likely to be less rigid.

Secondly all the methods assume a target variable and the goal of the discovery is to identify local direct causal relationships around the target variable, i.e. to look for direct causes or direct effects of the target variable. Although it would be desirable to construct a complete network of causal relationships among all the variables under consideration, it is infeasible to achieve this goal when there are a large number of variables. On the other hand, local causal discovery has high practical importance because an investigation often starts with exploring the direct causes of a known effect, e.g. lung cancer, or a direct effect of a causal factor such as exposure to a certain radiation. At the first stage of the investigation, we are more interested in the direct causes of lung cancer, such as smoking, instead of the causes for people to smoke. A causal relationships discovered using the methods presented in this book may be in either direction, i.e. the variable identified to be causally related to the target variable may be a cause or an effect of the target. However, finding causal relationships in either direction is very useful. For example, for a disease (as the target variable), knowing its causes and symptoms (effects) are both useful, while the former helps with treatment and the latter can assist with diagnosis. When we have already identified a causal relationship around a target variable, its direction can be determined with less difficulties as at this stage prior knowledge of the related variables or controlled experiments can be more easily applied.

Additionally, the two causal rule discovery methods are able to detect combined causes. Events can happen due to one single reason, but there are also cases where a result holds only when multiple factors work together to lead to it. For example, Internet overuse and low socioeconomic status individually may not be a cause of low academic performance of students, but when they are combined, they may affect students academic performance.

1.4 Limitations and Strength

It is arguable whether causality is definable or discoverable in data. We do not try to define causality in this book or to prove that the discovered relationships are causal. It is worthy of noting that the research endeavour in the past decades has made significant progress in the theory of causal discovery [13, 20], which has shown promising applications in classification and prediction [1, 2, 8]. The causal models, on which the methods in this book are based, are theoretically sound, however there is no guarantee that the discoveries by one of the methods are causal because (1) a data set normally does not satisfy the assumptions used for a causal definition that a method is based on; and (2) there is no analysis of the time sequence of events or the directions of the relationships.

So why do we need causal relationship discovery in addition to normal relationship discovery in data mining and machine learning literature?

Causal discovery is a step forward in data exploration and prediction. Let us look at association rule mining, which is a popular data mining method for relationship discovery. An association rule mining method normally generates a large number of association rules, for example hundreds of thousands to millions, most of which are spurious and redundant. Although many of the association rules can be removed by statistical analysis and redundancy analysis [10, 21], the remainder are still association rules which may be accidental, spurious and conditional as indicated in the previous discussions. Persistence is the main characteristics of a causal relationship, and the test of a causal relationship involves all other variables of a data set and considers all circumstances. Therefore causal relationships are less likely spurious or volatile than associations.

Let us look at another example. Decision tree is a popular machine learning method for relationship exploration [15]. The dependence of a variable on the class variable (measured by information gain) is the main criterion for selecting a variable to make a decision. Direct dependence and indirect dependence (intermediated by other variables) are not differentiated. For classification, such a differentiation makes no difference since trees built on direct dependence or indirect dependence very likely lead to the same classification. However, if we use the decision trees to intervene a real world process, direct dependence may produce an effect while indirect dependence may not. For example, when two variables, A and B, are dependent due to the effect of another variable C, in data we observe that (A and B) and (C and B) are correlated respectively. For classifying $B = 1$ and $B = 0$ in a data set, it does not matter which correlation to use since both give the same accuracy. However, manipulating A will not change B while manipulating C will. So whether a discovered relationship is causal or not makes a big difference in a real world intervention.

Causal relationships are more interpretable and actionable than those identified by normal relationship discovery methods. This book introduces four practical methods for causal discovery. The authors hope that the methods enable practitioners to use causal discovery in solving real life problems. Furthermore, as the causal discovery methods consider all other variables when testing the relationship between two variables, they have high demands on computational power. Therefore another aim of the book is to motivate researchers to design more efficient causal discovery methods.

References

1. C. F. Aliferis, A. Statnikov, I. Tsamardinos, S. Mani, and X. D. Koutsoukos. Local causal and Markov blanket induction for causal discovery and feature selection for classification Part I: Algorithms and empirical evaluation. *Journal of Machine Learning Research*, 11:171–234, 2010.
2. C. F. Aliferis, A. Statnikov, I. Tsamardinos, S. Mani, and X. D. Koutsoukos. Local causal and Markov blanket induction for causal discovery and feature selection for classification Part II: Analysis and extensions. *Journal of Machine Learning Research*, 11:235–284, 2010.
3. D. Clayton M. A. Hernan and N. Keiding. The Simpson's paradox unraveled. *International Journal of Epidemiology*, 40(3):780–785, 2011.
4. E. A. Hammel P. J. Bickel and J. W. O'Connell. Sex bias in graduate admissions: Data from berkeley. *Science*, 187(4175):398–404, 1975.
5. A. B. Hill. The environment and disease: Association or causation? *Proceedings of the Royal Society of Medicine*, 58:295–300, 1965.
6. A. R. Jadad and M. W. Enkin. *Randomized Controlled Trials Questions, Answers, and Musings*. Blackwell Publishing, 2nd. edition, 2007.
7. Z. Jin, J. Li, L. Liu, T. D. Le, B. Sun, and R. Wang. Discovery of causal rules using partial association. In *Data Mining (ICDM), 2012 IEEE 12th International Conference on*, pages 309–318, 2012.

8. M. Kalisch P. Büehlmann and M. H. Maathuis. Variable selection for high-dimensional linear models: partially faithful distributions and the PC-simple algorithm. *Biometrika*, 97:261–278, 2010.
9. J. Li, T. D. Le, L. Liu, J. Liu, Z. Jin, and B. Sun. Mining causal association rules. In *ICDM Workshops*, pages 114–123, 2013.
10. J. Li, J. Liu, H. Toivonen, K. Satou, Y. Sun, and B. Sun. Discovering statistically non-redundant subgroups. *Knowledge-Based Systems*, 67:315–327, 2014.
11. G. Malinas and J. Bigelow. Simpson's paradox. In Edward N. Zalta, editor, *The Stanford Encyclopedia of Philosophy*. Winter 2012 edition, 2012.
12. R. Matthews. Storks deliver babies (p= 0.008). *Teaching Statistics*, 22(2):1467–9639, 2000.
13. J. Pearl. *Causality: Models, Reasoning, and Inference*. Cambridge University Press, 2000.
14. H. Przyrembel T. Hfer and S. Verleger. New evidence for the theory of the stork. *Paediatric and Perinatal Epidemiology*, 18(1):88–92, 2004.
15. J. R. Quinlan. *C4. 5: programs for machine learning*. Morgan Kaufmann Publishers Inc., San Francisco, CA, USA, 1993.
16. M. L. Samuels. Simpson's paradox and related phenomena. *Journal of the American Statistical Association*, 88(421):81–88, 1993.
17. B. Sax. *The Mythical Zoo: An Encyclopedia of Animals in World Myth, Legend, and Literature*. ABC-CLIO, Inc, 2001.
18. W. R. Shadish, T. D. Thomas, and D. T. Campbell. *Experimental and Quasi-Experimental Designs for Generalized Causal Inference*. Houghton Mifflin, Boston, 2nd. edition, 2002.
19. B. Sibbald and M. Roland. Understanding controlled trials: Why are randomised controlled trials important? *BMJ*, 316(7126):201, 1 1998.
20. P. Spirtes, C. C. Glymour, and R. Scheines. *Causation, Predication, and Search*. The MIT Press, 2nd. edition, 2000.
21. G. I. Webb. Discovering significant patterns. *Machine Learning*, 71:1–31, 2009.

Chapter 2
Local Causal Discovery with a Simple PC Algorithm

Abstract This chapter presents the PC-simple algorithm and illustrates how to use the algorithm in the exploration for local causal relationships around a target variable. PC-simple is a simplified version of the PC algorithm, a classic method for learning a complete casual Bayesian network. We firstly discuss how the PC algorithm establishes causal relationships by the way of detecting persistent associations, then we introduce PC-simple in detail, followed by the discussions on PC-simple. The last section of this chapter introduces the R implementation of PC-simple.

2.1 Introduction

Bayesian networks [6, 8], a type of probabilistic graphical models [4], are a major means of causal representation and inference. Given the set of variables representing a domain, a Bayesian network presents the full joint probability of the variables by a directed acyclic graph (DAG) containing nodes representing the variables and arcs indicating dependence relationships between the variables.

More formally, with a set of variables, $\mathbf{V}$, a Bayesian network consists of a structure, the DAG, $\mathbf{G} = (\mathbf{V}, \mathbf{E})$, and the joint probability distribution $P(\mathbf{V})$ such that the Markov condition holds, i.e. for each node $X \in \mathbf{V}$, given its parent nodes $\mathbf{P}_a(X) \subset \mathbf{V}$, X is independent of all of its non-descendant nodes according to $P(\mathbf{V})$. Hence a Bayesian network provides a graphical representation of the conditional independence relationships among all the variables in $\mathbf{V}$, and as a result of the Markov condition the full joint probability distribution $P(\mathbf{V})$ can be factored into $P(\mathbf{V}) = \prod_{X \in \mathbf{V}} P(X|\mathbf{P}_a(X))$.

In the exemplar Bayesian network in Fig. 2.1, conditioning on an empty set of parents, A is independent of all its non-descendents, D, E and H; given A, variables B and C are independent of each other and they both are independent of D, E and H; D and E both have no parents, apart from being independent of each other they both are independent of A, B, C, Z and H; conditioning on its parents D, E, and Z,

J. Li et al., *Practical Approaches to Causal Relationship Exploration*,
SpringerBriefs in Electrical and Computer Engineering, DOI 10.1007/978-3-319-14433-7_2

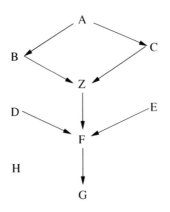

Fig. 2.1: An example Bayesian network

F is independent of A, B, C, and H; G is independent of any other variable given its parent F; H is isolated so it is independent of all other nodes; and finally Z is independent of A, D, E and H given its parents B and C. Furthermore, let $\mathbf{V}$ be the set of variables in the Bayesian network, the joint probability distribution $P(\mathbf{V}) = P(A)P(B|A)P(C|A)P(D)P(E)P(F|D,E,Z)P(G|F)P(H)P(Z|B,C)$.

A causal Bayesian network is a Bayesian network when its structure is considered as a causal DAG, i.e. an edge $X \to Y$ in the DAG represents that X is a direct cause of Y. When the set of causal assumptions (discussed in Sect. 2.3.3) are made to link causal relationships and probability distributions, we can learn the causal structures from observational data.

We can discover causal relationships in a data set by learning the structure of a causal Bayesion network, but we may not know the directions of the causal relationships since the orientation of edges in a Bayesian network may not be completely determined only based on the data set (Please read Sect. 2.3.3 for detailed discussions). Therefore by learning a Bayesian network from a data set, we can conclude that two variables have a causal relationship but we may not know which variable is the cause and which variable is the effect.

The PC algorithm [5, 8] is a commonly used method for learning the structure of a causal Bayesian network. With a data set, for a pair of variables or nodes (X,Y), the PC algorithm tests their conditional independence given the other variables, and it claims the non-existence of a causal relationship between X and Y, i.e. no edge to be drawn between X and Y, once it finds that X and Y are independent given some other variables. In other words, to determine whether there exists a persistent association between X and Y, the PC algorithm tests the association conditioning on all sub sets of all variables other than X and Y. The relationship is considered as causal only when the association exists given each of the conditioning sets.

Let us consider an exemplar binary data set containing attributes, Gender, College education, High school education, Manager, Clerk, and Salary, where the values of the variables are {*male, female*} for Gender, {*high, low*} for Salary, and {*yes, no*} for all other variables. Note that we use the term *target variable* to represent the outcome or effect, and *predictor variables* to represent the inputs or potential causes in this book. For example, variable Salary in this example is the target variable whose value is affected by other variables, and the remaining variables are predictor variables representing possible causes for having a high/low salary. To find out if Gender and Salary have a causal relationship, the independence between Gender and Salary is tested conditioning on each of the subsets of the other variables {*College education, High school education, Manager, Clerk*}, including the empty set. In the worst case, for each predictor variable, the number of conditional independence tests is 2^{m-1} where m is the number of predictor variables. Therefore, the PC algorithm only works on a data set with a small number of predictor variables.

In its algorithmic description, PC starts with a complete (undirected) graph with all the variables, $\mathbf{V}$, and removes the edge between a pair of nodes X and Y immediately after a subset $\mathbf{S} \subseteq \mathbf{V} \setminus \{X,Y\}$ is found such that X and Y are independent given $\mathbf{S}$. In order to reduce the number of conditional independence tests, the PC algorithm searches for the conditioning set for a pair of nodes in a level by level manner, i.e. searching the conditioning sets with $k+1$ variables only when the search of all size k conditioning sets fails.

PC-simple [2] was originally developed for efficient variable selection in high dimensional linear regression models. The algorithm produces a set of variables which have strong influence on the target variable. Interestingly this algorithm turned out to be a simplified version of the PC algorithm. This possibly explains why the implementation of PC-simple by its authors (named as PC-select) is included in *pcalg* [3], the R-package of the methods for causal inference with graphical models.

PC-simple utilises the same idea as the PC algorithm to detect persistent associations, and it also conducts a level-wise search for the conditioning set for a pair of nodes. However, instead of learning a causal DAG that captures all the causal relationships among all the given variables, PC-simple can be used to discover the local relationships around a given response or target variable. That is, given the data set $\mathbf{D}$ for a set of variables $(X_1, X_2, \ldots, X_m, Z)$ where Z is the target variable, PC-simple identifies all the possible direct causes and effects of Z, i.e. all Z's parents and children if we use the terminology for DAGs. In this case, it may be more meaningful to interpret the acronym "PC" in PC-simple as Parents and Children, instead of the names of the inventors (Peter Spirtes and Clark Glymour) of the PC algorithm.

With the goal of local relationship discovery, the algorithmic design of PC-simple, although follows the basic procedure of the PC algorithm, it has been simplified as presented in the next section.

2.2 The PC-simple Algorithm

Before describing the PC-simple algorithm, we firstly formally define conditional independence as follows.

Definition 2.1 (Conditional independence). Two variables X and Z are conditionally independent given a set of variables $\mathbf{S}$, denoted as $\mathrm{Ind}(X,Z|\mathbf{S})$, if $P(X = x, Z =$

$z|\mathbf{S} = s) = P(X = x|\mathbf{S} = s)P(Z = z|\mathbf{S} = s)$ for all the values of X, Z, and $\mathbf{S}$, such that $P(\mathbf{S} = s) > 0$. The cardinality or size of $\mathbf{S}$, denoted as $|\mathbf{S}|$, is known as the order of the conditional independence.

Given a data set $\mathbf{D}$ for a set of variables, the conditional independence between any two of the variables given any other variables can be tested using the data set at a specified significance level. Details of the commonly used conditional independence test methods are provided in Appendix A.

As shown in Algorithm 2.1 PC-simple takes as input the data set $\mathbf{D}$ for predictor variables $X_1, X_2, \ldots, X_m$ and the target variable Z, and produces $\mathbf{PC}$, the set of parents and children of Z. The input parameter for the algorithm is the significance level used for conditional independence tests.

Initially the $\mathbf{PC}$ set (i.e. $\mathbf{PC}^0$) contains all the predictor variables (Line 2 of Algorithm 2.1), and then PC-simple removes from the $\mathbf{PC}$ set those variables that are not Z's parents or children via conditional independence tests. The tests are done level by level of the cardinality of the conditioning sets, starting with an empty conditioning set (i.e. order zero conditional tests are done first). Each iteration of the **while** loop (Lines 3–12) generates $\mathbf{PC}^k$ from $\mathbf{PC}^{k-1}$ with the removal of variables that are independent of Z conditioning on $k-1$ variables in $\mathbf{PC}^{k-1}$.

Specifically, in the first iteration, PC-simple initially lets $\mathbf{PC}^1$ equal to $\mathbf{PC}^0$ (Line 5 with $k = 1$), then it checks through $\mathbf{PC}^0$ (Lines 6 to 11 with $k = 1$), and if a variable in $\mathbf{PC}^0$ is independent of Z given an empty set (note that in Line 7, $|\mathbf{S}| = 0$ when $k = 1$), $\mathbf{PC}^1$ is updated by removing the independent variable from it (Lines 8 and 9). At the end of the first iteration, $\mathbf{PC}^1$ contains only the variables that are associated with Z. In the second iteration, PC-simple initially lets $\mathbf{PC}^2$ equal to $\mathbf{PC}^1$, then it updates $\mathbf{PC}^2$ by removing from it the variables that are independent of Z given any other single variable in $\mathbf{PC}^1$. Similarly in the third iteration, $\mathbf{PC}^3$ is initially equal to $\mathbf{PC}^2$, then $\mathbf{PC}^3$ is updated by removing from it the variables that are independent of Z given any other two variables in $\mathbf{PC}^2$. This process iterates until the number of variables in $\mathbf{PC}^k$ is not more than k, and $\mathbf{PC}^k$ is output as the final $\mathbf{PC}$ set.

In the following, we use an example to run through the PC-simple algorithm.

Example 2.1. Supposing that we have a data set for variables $\{A,B,C,D,E,F,G,H\}$ and the target variable Z, such that the results of the conditional independence tests based on the data set can be represented by the Bayesian network structure in Fig. 2.1. Then the steps of PC-simple are given as the following.

1. $k = 0$ and $\mathbf{PC}^0 = \{A,B,C,D,E,F,G,H\}$.
2. As $|\mathbf{PC}^0| > 0$, enter the **while** loop.
3. $k = 1$ and $\mathbf{PC}^1 = \mathbf{PC}^0 = \{A,B,C,D,E,F,G,H\}$.
4. For each variable in $\mathbf{PC}^1$, order zero independence test with Z is conducted given $\mathbf{S} = \emptyset$. At the end of this iteration of the **while** loop, D, E and H are removed from $\mathbf{PC}^1$ since they are independent of Z, and $\mathbf{PC}^1 = \{A,B,C,F,G\}$.
5. $|\mathbf{PC}^1| > k$ $(k = 1)$, start the second iteration of the **while** loop.
6. $k = 2$ and $\mathbf{PC}^2 = \mathbf{PC}^1 = \{A,B,C,F,G\}$.
7. For each of the five variables in $\mathbf{PC}^2$, order one independence tests are conducted conditioning on any other single variable in $\mathbf{PC}^1$ each time $(|\mathbf{S}| = 1)$. At the end

---

Algorithm 2.1: The PC-simple algorithm [2]

Input: **D**, a data set for the set of predictor variables $\mathbf{X} = \{X_1, X_2, \dots, X_m\}$ and the target variable Z; and α, significance level for conditional independence tests.

Output: **PC**, the subset of $\{X_1, X_2, \dots, X_m\}$ that comprises parents and children of Z

1: let $k = 0$
2: let $\mathbf{PC}^k = \{X_1, X_2, \dots, X_m\}$
3: **while** $|\mathbf{PC}^k| > k$ **do**
4:   let $k = k + 1$
5:   let $\mathbf{PC}^k = \mathbf{PC}^{k-1}$
6:   **for** each $X \in \mathbf{PC}^{k-1}$ **do**
7:     **for** each $\mathbf{S} \in \mathbf{PC}^{k-1} \setminus \{X\}$ and $|\mathbf{S}| = k - 1$ **do**
8:       if X and Z are independent given $\mathbf{S}$ at significance level α
9:         let $\mathbf{PC}^k = \mathbf{PC}^k \setminus \{X\}$
10:     **end for**
11:   **end for**
12: **end while**
13: output $\mathbf{PC}^k$

---

of this iteration, only G is removed from the **PC** set and $\mathbf{PC}^2 = \{A, B, C, F\}$ (see Fig. 2.1 which shows that G is independent of Z given its parent F, which blocks G and Z).

8. $|\mathbf{PC}^2| > k$ $(k = 2)$, start the third iteration of the **while** loop.
9. $k = 3$ and $\mathbf{PC}^3 = \mathbf{PC}^2 = \{A, B, C, F\}$.
10. For each of the four variables, its independence with Z is tested given the combination of any other two variables in $\mathbf{PC}^2$. Then at the end of this iteration, A is removed as Z is independent of A given its two parents B and C, and $\mathbf{PC}^3 = \{B, C, F\}$.
11. Since $|\mathbf{PC}^3| = k$ $(k = 3)$, the **while** loop exits, and $\mathbf{PC}^3$ is output as the final **PC** set of Z. $\mathbf{PC} = \{B, C, F\}$.

That is, the **PC** set of Z is $\{B, C, F\}$, which is consistent with the structure in Figure 2.1.

2.3 Discussions

2.3.1 Complexity of PC-simple

Referring to Algorithm 2.1, the complexity of PC-simple comes from the time taken for conditional independence tests. In the worst case, i.e. all the variables $X_1, X_2, \dots, X_m$ are parents or children of Z, the maximum order of conditional independence tests is $m - 1$ (starting with order zero test); at level k $(0 \le k \le m - 1)$, all the m variables in $\{X_1, X_2, \dots, X_m\}$ need to be tested for conditional independence with Z; and for each of the variables, there are up to $\binom{m-1}{k}$ conditioning sets to be tested. Therefore overall, in the worst case, the number of conditional independence tests is $m\sum_{k=0}^{m-1}\binom{m-1}{k}$, which is $m2^{m-1}$. Hence in the worst case, the time complexity of PC-simple is exponential to the number of variables, which means that when the number of variables or m increases, the performance of PC-simple may degrade dramatically.

However, if most of the variables in $\{X_1, X_2, \ldots, X_m\}$ are not direct causes or effects of Z (or the degree of Z is small in an underlying Bayesian network), the size of the **PC** set is reduced quickly, and the same for the number of conditioning sets in each iteration of the **while** loop. So the **while** loop will be completed after a relatively small number of iterations because $(|\mathbf{PC}^k| > k)$ is violated.

Hence when the number of direct causes and effects of a target is small, PC-simple can handle high dimensional data sets. For example, in [2] PC-simple was applied to a real world data set with over 4000 variables. Such data sets are called sparse data sets since the underlying Bayesian networks are sparse (the degree of a node is very small). Normally, when the number of causal variables is 30 or more, PC-simple will not work well on a normal desktop computer.

2.3.2 False Discoveries of PC-simple

There are two sources for the false discoveries of PC-simple (1) algorithm design; and (2) input data.

Referring to Algorithm 2.1, PC-simple updates the **PC** set of Z after order k conditional independence tests are done, by removing all the variables that have been tested to be conditionally independent of Z. This update is reasonable in terms of removing conditionally independent variables of Z as a result of order k conditional independence tests. However, as the independence tests at the next level will be conditioned on the subsets of the updated **PC** set, the removed variables will not be considered as part of the conditioning sets, which will lead to false discoveries.

For example, supposing Fig. 2.2 shows the underlying causal relationships among the nine variables, where the true **PC** set of the target variable Z is $\{B, C, F\}$. However, the discovered **PC** set by PC-simple will also include G (assuming all the conditional independence tests are correct), which is a false positive. This is caused by the removal of E based on the result of order zero conditional independence test, as E is independent of Z (given an empty set). So E will not be in any of the conditioning sets for higher order conditional independence tests, and as a result, G will not be removed since G is independent of Z given both E and F (G's parents). G is included in the final **PC** list of Z as a false positive.

In [1], the idea of a symmetry correction is introduced to remove such false positives in local causal discovery. If G is a true parent or a child node of Z, then Z should be in the **PC** set of G. Let $\mathbf{PC}(Z)$ and $\mathbf{PC}(G)$ be the **PC** sets of Z and G respectively. In the example, PC-simple outputs $\mathbf{PC}(Z) = \{B, C, F, G\}$, and $\mathbf{PC}(G) = \{E, F\}$. Z is not in $\mathbf{PC}(G)$. Based on the symmetry property, G should not be in $\mathbf{PC}(Z)$ and hence be removed. The symmetry correction, however, introduces higher time complexity because we need to test the symmetry for each variable in $\mathbf{PC}(Z)$ in order to

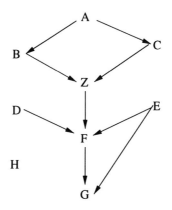

Fig. 2.2: An example scenario for false discovery of PC-simple

remove the false positives. The studies in [1] with a number of real world data sets have shown that the situations in which the false positives may occur are not often in practice.

PC-simple will produce false discoveries if a conditional independence test is incorrect at any stage. An incorrect test causes a variable to be falsely removed or included, and such a false exclusion or inclusion will result in incorrect conditioning sets for conditional independence tests at the following levels. As a consequence, an incorrect test result may cause a chain of false positives and/or false negatives. Therefore the algorithms should be used with caution if there is no sufficient number of samples or there is selection bias in the collection of samples. A number of different conditional independent tests should be used to ensure the reliability of the results.

2.3.3 The Causal Assumptions

PC-simple is based on causal Bayesian network theory, and it follows the same causal assumptions as those by causal Bayesian networks. In the context of causal Bayesian networks, to link a causal graph with a probability distribution, the two essential causal assumptions are the Causal Markov Condition and the Faithfulness Condition, as described below.

Let $G = (V, E)$ be a directed acyclic graph (DAG), where the nodes V represent a set of random variables and the edges E represent causal relationships between the nodes. An edge from a node X to a node Y indicates that X is a direct cause of Y, and X is known as the parent of Y. Let $P(V)$ be the joint probability distribution of V.

Assumption 2.1 (Causal Markov Condition [8]) *G and P(V) satisfy the Causal Markov Condition if and only if given the set of all its parents, a node of G is independent of all its non-descendents according to P(V).*

Simply speaking, this condition states that every edge in a DAG implies a probabilistic dependence. In other words, this condition has drawn a one way link from edges to dependencies.

However, it may be possible that some of the conditional independence relationships are not reflected by the DAG, so we need the following assumption to draw the link from dependencies to edges.

Assumption 2.2 (Faithfulness Condition [8]) *G and $P(V)$ satisfy the Faithfulness Condition if and only if every conditional independence relationship in $P(V)$, $\text{Ind}(X,Z|S)$ is reflected in G in the way that S is the set of all parents of X (or Z) and Z (or X) is not a descendent or a parent of X (or Z).*

In other words, Faithfulness Condition assumes that if two variables are probabilistically dependent, there must be a corresponding edge between the two variables in the graph.

With the above two conditions, we have established the mapping between conditional dependence relationships specified by $P(\mathbf{V})$ and the causal relationships (edges) represented by a causal DAG. Therefore, an algorithm that discovers the correct conditional independence relationships also produces the mapped true causal relationships.

There may be more than one causal DAGs which faithfully reflect the same conditional independence relationships in a probability distribution. However, these DAGs have the same and unique underlying undirected graph (skeleton), although the directions of edges in the DAGs may be different [5]. For example, the DAGs $A \rightarrow B$ and $A \leftarrow B$ reflect the same conditional independence relationship $\neg Ind(A, B|\emptyset)$, i.e. A and B are dependent. The skeleton of both DAGs is $A - B$.

In causal Bayesian network structure learning, an algorithm is able to obtain from the data (with sufficient number of samples) an equivalence class of the causal DAGs [5]. The equivalence class can be represented as a partially directed graph, where a directed edge $X \rightarrow Y$ indicates that all causal DAGs in the class have the edge $X \rightarrow Y$ and an undirected edge $X - Y$ shows that some of the causal DAGs in the class contain $X \rightarrow Y$ while some of them contain $X \leftarrow Y$.

The set of parents and children nodes of a node in a learned causal Bayesian network are deterministic although the direction of the edge between a node in the set and the target node may be nondeterministic. Local causal discovery algorithms like PC-simple and HITON-PC are able to learn a unique parents and children set of Z in a data set with sufficient number of samples under the same causal assumptions. The cause-effect relationships around a given target variable Z are not directed, i.e. we do not know whether a variable in the learned set is the cause or effect of the target variable. Nevertheless, the undirected causal relationships are useful for causal exploration as discussed in Chap. 1.

Here we indeed have also assumed the following causal sufficiency about the data sets used by the algorithms.

Assumption 2.3 (Causal sufficiency [8]) *For every pair of variables which have their observed values in a given data set, all their common causes also have observations in the data set.*

This assumption assumes that there is no unmeasured or hidden causes. For example, if a common cause of two variables is unobserved in a data set, PC-simple may conclude that the two variables have a causal relationship while in fact they do not.

2.4 An Implementation of PC-simple

PC-simple [2] is implemented in *pcalg* [3] as the *pcSelect* function. In this section, we present some examples in *R* [7] to demonstrate the usage of the *pcSelect* function. We assume that readers are familiar with *R*. If not, the *R* introduction documentation can be found in [9].

2.4.1 Example 1: Using PC-simple in pcalg

In this example, we use the built-in data set, *gmB*, in the *pcalg* package for demonstrating the usage of the *pcSelect* function. The *gmB* data set includes five binary variables (columns) with 5000 samples (rows). The data is stored in *gmB$x* and the known true DAG is *gmB$g* as shown in Fig. 2.3. In the following, we assume that variable *V2* (i.e. node 2 in Fig. 2.3) is the target variable and other variables are predictor variables, and we apply the PC-simple algorithm to the *gmB* data set.

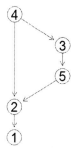

Fig. 2.3: The ground truth Bayesian network of the *gmB* data set

```
> library(pcalg)
> library(Rgraphviz) # for drawing graphs
> data(gmB)
> plot(gmB$g)
> results = pcSelect(gmB$x[,2], gmB$x[,-2], alpha=0.01)
> results
```

```
$G
    V1      V3      V4      V5
  TRUE  FALSE    TRUE    TRUE
$zMin
[1] 29.4191620   0.6714314   5.5774279  44.3422923
```

The result ($G) demonstrates that variables V1, V4, V5 are in the parents and children set of the target variable, while variable V3 is not. zMin is the set of *z*-values of the conditional independence tests corresponding to each of the predictor variables. The bigger a *z*-value is, the stronger the association the predictor variable and the target variable have. More details on the conditional independence test used in PC-simple can be found in Appendix A.

2.4.2 Example 2: Using the Data Sets of Figs. 2.1 and 2.2

The data set that has the same dependence relationships as the Bayesian network shown in Fig. 2.1 can be downloaded from:

http://nugget.unisa.edu.au/Causalbook/

The data set contains eight predictor variables and the target variable as described in Example 2.1. The data set was generated using the TETRAD software downloaded from *http://www.phil.cmu.edu/tetrad/*. Assume that the data set *Example21.csv* has been downloaded and stored in the *R* working directory. We now use the *pcSelect* function to find the parents and children set of *Z*.

```
> library(pcalg)
> data=read.csv("Example21.csv", header=TRUE, sep=",")
> pcSimple.Z=pcSelect(data[,9], data[,-9], alpha=0.01)
> pcSimple.Z
$G
     A      B      C      D      E      F      G      H
 FALSE   TRUE   TRUE  FALSE  FALSE   TRUE  FALSE  FALSE
$zMin
[1]   1.85790685  144.57474808   16.65675448   0.27538423
[5]   0.76391512   15.94735566    0.07287629   0.23537966
```

The parents and children set of *Z* is {*B,C,F*}, which is consistent with the graph in Example 2.1. We can also verify this result against the global causal structure (Fig. 2.4) learned by the PC algorithm [5, 8]. The DAGs in Figs. 2.1 and 2.4 have the same skeleton and *Z* has the same **PC** set in both Figures. The following codes are used to learn the causal structure from the *Example21.csv* data set using the PC algorithm.

```
> pc.example=pc(suffStat=list(dm=data, adaptDF=FALSE),
  indepTest=binCItest,alpha=0.01,labels=colnames(data))
> library(Rgraphviz)
> plot(pc.example)
```

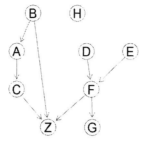

Fig. 2.4: The causal structure learned by PC algorithm for the data set in Example 2.1

We now use the data set that has the same dependence relationships as the Bayesian network shown in Fig. 2.2, which can be downloaded from:

http://nugget.unisa.edu.au/Causalbook/

This is a scenario where PC-simple algorithm will have false discoveries. We use the *pcSelect* function to learn the local causal structure around Z.

```
> library(pcalg)
> data=read.csv("Example22.csv", header=TRUE, sep=",")
> pcSimple.Z = pcSelect(data[,9], data[,-9], alpha=0.01)
> pcSimple.Z
$G
      A       B      C      D      E       F       G      H
   TRUE    TRUE  FALSE  FALSE  FALSE   TRUE    TRUE  FALSE
$zMin
[1]    0.8534152    9.1261327 109.5619224    1.2201710
[5]    0.6406870   96.6218472   3.0954975    1.1304557
```

We can see from the result that the parents and children set of Z is $\{A,B,F,G\}$, and thus the result involves a false discovery, G, as discussed in Sect. 2.3.2. The false discovery can also be confirmed by comparing the result with the global causal structure learned by the PC algorithm. The codes for learning the global causal structure with the *Example22.csv* data set are in the following, and the causal structure is shown in Fig. 2.5.

```
> pc.example22=pc(suffStat=list(dm=data, adaptDF=FALSE),
  indepTest=binCItest, alpha=0.01, labels=colnames(data))
> library(Rgraphviz)
> plot(pc.example22)
```

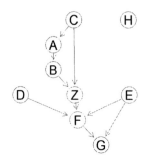

Fig. 2.5: The causal structure learned by PC algorithm for the data set of Fig. 2.2

The DAGs in Figs. 2.5 and 2.2 have the same skeleton and Z has the same **PC** set, $\{A, B, F\}$, in both Figures. Therefore, the result from PC-simple involves a false discovery, G. However, the false discovery can be removed if we conduct the symmetry correction as discussed in Sect. 2.3.2. The parents and children set of G can be found using the following codes:

```
> pcSimple.G = pcSelect(data[,7], data[,-7], alpha=0.01)
> pcSimple.G
$G
      A      B      C      D      E      F      H      Z
  FALSE  FALSE  FALSE   TRUE   TRUE   TRUE  FALSE  FALSE
$zMin
[1]   0.660409975   0.008014493   2.517720119   3.215486235
[5]  34.630702580  33.470052040   0.020871355   2.359975957
```

We can observe that Z is not in the parents and children set of Z, and therefore based on the symmetry property, G can be removed from the parents and children set of Z.

References

1. C. F. Aliferis, A. Statnikov, I. Tsamardinos, S. Mani, and X. D. Koutsoukos. Local causal and Markov blanket induction for causal discovery and feature selection for classification Part I: Algorithms and empirical evaluation. *Journal of Machine Learning Research*, 11:171–234, 2010.
2. M. Kalisch P. Büehlmann and M. H. Maathuis. Variable selection for high-dimensional linear models: partially faithful distributions and the PC-simple algorithm. *Biometrika*, 97:261–278, 2010.
3. M. Kalisch, M. Mächler, D. Colombo, M. H. Maathuis, and P. Bühlmann. Causal inference using graphical models with the R package pcalg. *Journal of Statistical Software*, 47(11):1–26, 5 2012.
4. D. Koller and N. Friedman. *Probabilistic Graphical Models: Principles and Techniques*. The MIT Press, 2009.
5. R. E. Neapolitan. *Learning Bayesian Networks*. Prentice Hall, 2003.

6. J. Pearl. *Causality: Models, Reasoning, and Inference.* Cambridge University Press, 2000.
7. R Core Team. *R: A Language and Environment for Statistical Computing.* R Foundation for Statistical Computing, Vienna, Austria, 2014.
8. P. Spirtes, C. C. Glymour, and R. Scheines. *Causation, Predication, and Search.* The MIT Press, 2nd. edition, 2000.
9. W. N. Venables, D. M Smith, R Development Core Team, et al. An introduction to R, 2002.

Chapter 3
A Local Causal Discovery Algorithm for High Dimensional Data

Abstract In this chapter we introduce HITON-PC, another local causal discovery algorithm for finding the parents and children of a given target variable. Similar to PC-simple, HITON-PC applies conditional independence tests to identify strong and persistent associations between variables, but with a different approach to pruning the search space for the tests. This chapter firstly presents the basic idea of HITON-PC, then the algorithm is described in detail and illustrated with a simple example. The time complexity and false discoveries of HITON-PC are also discussed, and the chapter ends with the introduction to a software tool containing the implementation of HITON-PC.

3.1 Introduction

The causal Markov condition and faithfulness assumption link up probabilistic independence/dependence and causal relationships encoded in a DAG, allowing sound causal discovery algorithms to be developed based on conditional independence tests (known as constraint-based algorithms). A key to the reliability of these algorithms is the complete coverage of conditional independence tests so that persistent associations can be identified correctly. That is, a causal relationship between two variables can be claimed if and only if the two variables are tested to be not independent given all subsets of other variables. However, conducting all the conditional independence tests is often computationally infeasible. Therefore a major objective of the design of constraint-based algorithms is to reduce the number of conditional independence tests while generating reliable results.

As we have seen from the previous chapter, PC-simple achieves this goal by conducting level or order wise conditional independence tests and shrinking the parents and children (**PC**) set at each level. Although the algorithm starts with a larger **PC** set (initially the full list of predictor variables), the **PC** set is updated at each level by removing from it variables that are independent of the target variable,

© The Author(s) 2015   23
J. Li et al., *Practical Approaches to Causal Relationship Exploration*,
SpringerBriefs in Electrical and Computer Engineering, DOI 10.1007/978-3-319-14433-7_3

and higher order tests will be done with a smaller **PC** set. However, only for data sets of sparse true causal relationships, it is possible to use the algorithm in practice.

In this chapter, we will introduce another constraint-based algorithm, HITON-PC for local causal discovery, specifically the semi-interleaved HITON-PC presented in [1]. For simplicity of presentation, we will call it HITON-PC throughout the book. This algorithm takes a rather different approach to reducing the number of conditional independence tests:

1. A priority queue is used to store all the variables associated with the target variable, in descending order of the strength of associations. Variables in the priority queue are candidates to be added to the **PC** set.
2. The **PC** set is built from an empty set, and is expanded gradually. For each expansion of the **PC** set, a candidate is taken from the queue and included in the **PC** set temporarily, and then the elimination strategy is applied to determine if this candidate is to be kept in **PC**, based on the results of the conditional independent tests with the target variable and given variables in the current **PC** set. Since variables having stronger associations with the target are tested earlier, it is expected that conditioning on them, non causal variables can be identified more quickly.
3. For the conditional independence tests, a threshold, max_k is used to specify the highest order of tests to be conducted. This is optional, but when it is set, the efficiency will be improved greatly.

HITON-PC has a more complicated algorithm design than PC-simple, but it has better scalability with variables than PC-simple, scaling up to the order of 10,000 variables, as the complexity has been reduced from exponential to polynomial. However, the quality of discoveries will be sacrificed slightly if the restriction on the maximum order of conditional independence tests is imposed.

3.2 The HITON-PC Algorithm

The pseudo code of HITON-PC is given in Algorithm 3.1. The input of the algorithm includes a data set for a set of predictor variables and a given target variable, and two parameters, one for limiting the order of conditional independence tests, and one for the significance level of the conditional independence tests. The output of the algorithm is the set of parents and children of the given target variable.

Initially the **PC** set is empty (Line 1), and the algorithm creates the **OPEN** list to hold all the predictor variables that are associated with the target Z (Line 2). Variables in **OPEN** are sorted in descending order according to the strength of associations, and a removal operation on the list will always takes out the first variable (which has the greatest association strength in the current **OPEN** list), so **OPEN** works as a priority queue.

The purpose of the ordering is to include the variables that are most likely to be parents or children in **PC** first so that they can be used to effectively prune the other

Algorithm 3.1: The HITON-PC algorithm [1]

Input: $\mathbf{D}$, a data set for the set of predictor variables $\mathbf{X} = \{X_1, X_2, \ldots, X_m\}$ and the target variable Z; max_k, threshold of order of conditional independence test; and α, significance level conditional independence test.
Output: $\mathbf{PC}$, the subset of $\mathbf{X}$ that comprises parents and children of Z

1: let $\mathbf{PC} = \emptyset$
2: let $\mathbf{OPEN}$ contain all variables associated with Z, sorted in descending order of strength of associations.
3: **while OPEN** $\neq \emptyset$ **do**
4:     remove the first variable X from $\mathbf{OPEN}$
5:     insert X to the end of $\mathbf{PC}$
6:     **for** each $\mathbf{S} \subseteq \mathbf{PC} \setminus \{X\}$ and $|\mathbf{S}| \leq max_k$ **do**
7:         **if** X and Z are independent given $\mathbf{S}$ at significance level α **then**
8:             remove X from $\mathbf{PC}$ and go to Line 3
9:         **end if**
10:    **end for**
11: **end while**
12: **for** each variable X in $\mathbf{PC}$ **do**
13:    **for** each $\mathbf{S} \subset \mathbf{PC} \setminus \{X\}$ and $\mathbf{S} \not\subset \mathbf{PC}_{<X}$, where $\mathbf{PC}_{<X}$ comprises the elements in $\mathbf{PC}$ that were added before X and $|\mathbf{S}| \leq max_k$ **do**
14:        **if** X and Z are independent given $\mathbf{S}$ at significance level α **then**
15:            remove X from $\mathbf{PC}$
16:        **end if**
17:    **end for**
18: **end for**
19: output $\mathbf{PC}$

variables early. HITON-PC makes use of the strength of the associations between predictor variables with Z as the criterion. Variables that are highly associated with Z will be ranked high. An alternative criterion is to use the strength of the conditional dependence between a variable and Z [4]. Given a pair of variables, there are a number of conditions for testing conditional dependence and the minimum dependence is used for the ordering. The work in [1, 2] indicates that the strength of association is a simple and effective criterion.

After the initialisation stage, the algorithm applies the inclusion and elimination strategies interleavingly to variables in $\mathbf{OPEN}$ to expand the $\mathbf{PC}$ set (Lines 3 to 11).

During each iteration of the **while** loop, the variable at the front of $\mathbf{OPEN}$ is removed and *included* in $\mathbf{PC}$ (Lines 4 and 5). Then the elimination step (Lines 6 to 10) immediately tests if the newly added variable X is independent of the target given the other variables in the current $\mathbf{PC}$ list. Once it is found that X is independent of the target given a subset of $\mathbf{PC}$ (excluding X), X is *eliminated* from $\mathbf{PC}$, and a new iteration starts. If conditioning on each of the subsets of $\mathbf{PC}$ with less than or equal to max_k variables, X is not independent of the target variable, X is maintained in $\mathbf{PC}$ for the moment.

When $\mathbf{OPEN}$ becomes empty, HITON-PC carries out the elimination step once more, but this time it checks each variable in $\mathbf{PC}$ to confirm its membership of $\mathbf{PC}$

(Lines 12 to 18). That is, for each variable X that is currently in **PC**, if **S**, a subset with no more than max_k variables of current **PC** (excluding X), can be found such that conditioning on **S**, X is independent of the target variable, X is eliminated from the **PC** set; if no such subset exists for X, then X stays in **PC**, permanently. Note that during this step, a conditioning set $\mathbf{S} \subset \mathbf{PC} \setminus \{X\}$, but $\mathbf{S} \not\subset \mathbf{PC}_{<X}$, where $\mathbf{PC}_{<X}$ comprises the elements in **PC** that were added before X, because in the previous step (Lines 6-10) the conditional independence between X and Z has been tested given the subsets of variables added to **PC** earlier than X.

After the check has been done for each variable, the confirmed **PC** gives us the final output of HITON-PC.

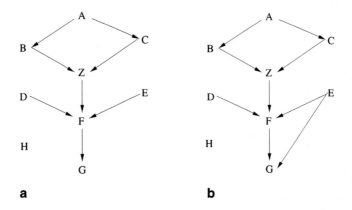

Fig. 3.1: **a** An exemplar Bayesian network to demonstrate the algorithm; **b** An exemplar Bayesian network to demonstrate false discoveries. Both are reproduced from Chap. 2 for the convenience of readers

In the following, we use an example to run through HITON-PC.

Example 3.1. Given a data set for the 8 predictor variables, A, B, C, D, E, F, G and H, and the target variable Z, assume that the data set indicates the conditional independence/dependence relationships shown in Fig. 3.1a. From the figure we see that D, E and H are not associated with Z. For the other five predictor variables, we assume the following (descending) order of the strength of their associations with Z: B, C, A, G, F.

Let $max_k = 2$, HITON-PC goes through the steps as described below. In the description, the variables in **OPEN** are enclosed in a pair of square brackets to indicate that they are listed in order of the strength of their associations with Z.

1. Initially **OPEN** $= [B, C, A, G, F]$ because H, D and E are not associated with Z, and **PC** $= \{\}$.
2. In the first iteration of the **while** loop:

   - inclusion: B is removed from **OPEN** and added to PC, so that **OPEN** $= [C, A, G, F]$ and **PC** $= \{B\}$;

- elimination: As $\mathbf{PC} \setminus \{B\}$ is empty, no conditional independence test is conducted with respect to B and the target, and B is kept in $\mathbf{PC}$.

3. In the second iteration:

- inclusion: C is removed from $\mathbf{OPEN}$ and added to $\mathbf{PC}$, we have $\mathbf{OPEN} = [A, G, F]$, and $\mathbf{PC} = \{B, C\}$;
- elimination: $\mathbf{PC} \setminus \{C\} = \{B\}$, so the test for $\mathrm{Ind}(C, Z|B)$ is conducted. Ind stands for conditional independence. As the result is false (see Fig. 3.1a), C is kept in $\mathbf{PC}$.

4. In the third iteration:

- inclusion: $\mathbf{OPEN} = [G, F]$ and $\mathbf{PC} = \{B, C, A\}$ after A is removed from $\mathbf{OPEN}$ and added to $\mathbf{PC}$
- elimination: since $\mathbf{PC} \setminus \{A\} = \{B, C\}$, the tests for $\mathrm{Ind}(A, Z|B)$, $\mathrm{Ind}(A, Z|C)$, and $\mathrm{Ind}(A, Z|B, C)$ are conducted. Because the last test returns true, A is eliminated from $\mathbf{PC}$, i.e. $\mathbf{PC} = \{B, C\}$.

5. In the fourth iteration:

- inclusion: G is removed from $\mathbf{OPEN}$ and added to $\mathbf{PC}$, which gives us $\mathbf{OPEN} = [F]$ and $\mathbf{PC} = \{B, C, G\}$;
- elimination: $\mathbf{PC} \setminus \{G\} = \{B, C\}$, the tests for $\mathrm{Ind}(G, Z|B)$, $\mathrm{Ind}(G, Z|C)$ and $\mathrm{Ind}(G, Z|B, C)$ are conducted, and all return false, so G is kept in $\mathbf{PC}$ and $\mathbf{PC} = \{B, C, G\}$.

6. In the fifth iteration:

- inclusion: F is taken from $\mathbf{OPEN}$ and added to $\mathbf{PC}$ so that $\mathbf{OPEN} = [\ ]$ and $\mathbf{PC} = \{B, C, G, F\}$;
- elimination: because $\mathbf{PC} \setminus \{F\} = \{B, C, G\}$ and $max_k = 2$, the tests for $\mathrm{Ind}(F, Z|B)$, $\mathrm{Ind}(F, Z|C)$, $\mathrm{Ind}(F, Z|D)$, $\mathrm{Ind}(F, Z|B, C)$, $\mathrm{Ind}(F, Z|B, G)$, $\mathrm{Ind}(F, Z|C, G)$ are conducted. No test returns true, so F is kept in $\mathbf{PC}$ and $\mathbf{PC} = \{B, C, G, F\}$.

7. As $\mathbf{OPEN}$ is now empty, the **while** loop is terminated.
8. The **for** loop is executed with the current variables in $\mathbf{PC}$ (at the start, $\mathbf{PC} = \{B, C, G, F\}$):

- For B, the tests of $\mathrm{Ind}(B, Z|C)$, $\mathrm{Ind}(B, Z|G)$, $\mathrm{Ind}(B, Z|F)$, $\mathrm{Ind}(B, Z|C, G)$, $\mathrm{Ind}(B, Z|C, F)$, $\mathrm{Ind}(B, Z|G, F)$ are conducted, and all return false, so B is permanently kept in $\mathbf{PC}$.
- For C, the tests of $\mathrm{Ind}(C, Z|G)$, $\mathrm{Ind}(C, Z|F)$, $\mathrm{Ind}(C, Z|B, G)$, $\mathrm{Ind}(C, Z|B, F)$, $\mathrm{Ind}(C, Z|G, F)$ are conducted, and all return false, so C is permanently kept in $\mathbf{PC}$. Note that $\mathrm{Ind}(C, Z|B)$ has been tested before in the second iteration.
- For G, the first test of $\mathrm{Ind}(G, Z|F)$ returns true. So G is permanently excluded from $\mathbf{PC}$, and $\mathbf{PC} = \{B, C, F\}$.
- For F, no conditional independence test is conducted since all variables in $\mathbf{PC} \setminus \{X\}$ were added before F, and the conditional independence between F and Z given the subsets of these variables have been tested in the fifth iteration. F is permanently kept in $\mathbf{PC}$.

9. The **for** loop ends, and $PC = \{B, C, F\}$, which is the output of the algorithm. From Fig. 3.1a, we can see that HITON-PC correctly identifies the parents and children set of Z.

3.3 Discussions

3.3.1 Complexity of HITON-PC

From Algorithm 3.1, the time taken by HTION-PC can be roughly divided into two parts, the time for initialisation (Line 2) and the time for the conditional independence tests conducted within the **while** loop and the **for** loop.

Initialising the **OPEN** list firstly requires the calculation of pair-wise association between each predictor variable and the target, which has the time complexity of $O(m)$, i.e. linear to the number of predictor variables, m. Sorting the associated variables with an efficient sorting algorithm, e.g. quick sort, will have time complexity of $O(m \log m)$, assuming that all predictor variables are associated with the target. So overall the complexity of the initialisation stage is $O(m \log m)$ in the worst case.

Comparing to the time taken by the conditional independence tests, the complexity of the initialisation stage is negligible. It is difficult to calculate the number of conditional independence tests precisely as in the **while** and the **for** loops, the size of the **PC** set constantly changes. In the worst case, i.e. when for each of the m predictor variables all conditional independence tests of order 1 up to $m-1$ are conducted, the total number of conditional independence tests is $m \sum_{k=1}^{m-1} \binom{m-1}{k}$, which is $O(m2^m)$, exponential to the number of variables.

In practice, however, HITON-PC has much lower complexity. Firstly this is due to the heuristics employed: (1) the use of the **OPEN** queue, which enables fast pruning of the search space of the parents and children variables; (2) a variable is eliminated based on the independence tests given only current members of the **PC** set. Therefore it is anticipated that in each iteration of the loops, the size of the temporary **PC** set is capped by the size of the true **PC** set, $|PC|$, so there will be at most $m \sum_{k=1}^{|PC|-1} \binom{|PC|-1}{k}$, i.e. $O(m2^{|PC|})$ conditional independence tests. As in real life, most high-dimensional problems are sparse, i.e. $|PC| \ll m$, thus $O(m2^{|PC|}) \ll O(m2^m)$. Secondly, since HITON-PC allows user to restrict the maximum order of independence tests, max_k, the number of the tests at most will be $m \sum_{k=1}^{max_k} \binom{|PC|-1}{k}$, i.e. $O(m|PC|^{max_k})$, which is polynomial to the number of variables.

Therefore, when the target has a small number of parents and children (i.e. the problem is sparse) and a small max_k is used, it is possible for HITON-PC to cope with high dimensional data efficiently.

3.3.2 False Discoveries of HITON-PC

Similar to PC-simple discussed in the previous chapter, HITON-PC also has two sources of false discoveries: algorithm design, and the input data and the reliability of the conditional independence tests.

As a result of the algorithm design, mainly two types false discoveries can happen.

Firstly, with HITON-PC, once a variable is eliminated from the **PC** set, the variable will not be used in the conditioning set for the tests for other variables. This may introduce false positives since it is only possible for some variables to be removed conditioning on the variables previously removed. For example, in Fig. 3.1b, E is removed in the initialisation step (Line 2) since it is independent of Z. In the future steps, G will not be removed since G is tested to be independent of Z conditioning on $\{E, F\}$. Because E has been removed, $\{E, F\}$ will not be a conditioning set for the test between G and Z, and hence G will be kept in **PC**, resulting in a false discovery. As discussed in [1], a symmetry correction can remove such false positives. Since the parent and child relationship of two nodes is symmetric, G is a parent (or child) of Z and Z will be a parent (or child) of G. When we set G as the target and find its parents and children set, if Z is not in this set then G should be removed from the parents and children set of Z. As a result, the false discovery caused by algorithm design is corrected. The symmetric correction involves high time complexity because we need to test the symmetry for each variable in the **PC** set.

Secondly, the restriction of the maximum order of conditional independence tests leads to false positives because lower order tests may not be able to detect all variables that are independent of the target variable. As a result, those undetected variables are incorrectly included in the **PC** set as false positives. This is a tradeoff of quality and efficiency.

The input data may also cause false discoveries. If a data set has insufficient number of samples or there is a selection bias in the collection of samples, the outcome of HITON-PC may include high proportion of false discoveries. In practice, statistical tests are prone to mistakes since data may not satisfy the distribution assumptions for a statistical test. It is recommended to use a number of different types of conditional independence tests for reliable results.

3.4 An Implementation of HITON-PC

An R implementation of HITON-PC is available as part of the *bnlearn* package [3]. We will go through some examples of using HITON-PC with *bnlearn* in the following.

3.4.1 *Example 1: Using HITON-PC in* **bnlearn**

The function *learn.nbr* in *bnlearn* is implemented to learn the local causal structure around a target node. This function can be used with different local causal structure learning algorithms, including HITON-PC. In this example, we use the built-in *asia* data set from the *bnlearn* package to demonstrate the usage of HITON-PC in local causal discovery. The *asia* data set contains eight binary variables, D (dyspnoea), T (tuberculosis), L (lung cancer), B (bronchitis), A (visit to Asia), S (smoking), X (chest X-ray), and E (tuberculosis versus lung cancer/bronchitis).

We firstly use the function *si.hiton.pc* for learning the global causal structure from the data set. The following codes show how to learn the global causal structure from the *asia* data set and the result is shown in Fig. 3.2.

```
> global.network = si.hiton.pc(asia, alpha=0.01)
> plot(global.network)
```

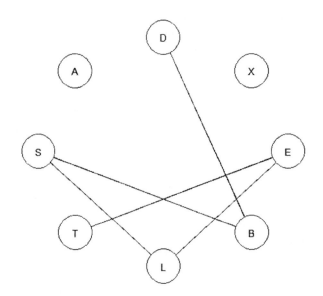

Fig. 3.2: The global causal structure of the *asia* data set

We now assume that node E is the target variable, and we apply HITON-PC to learn the parents and children set of E.

```
> library(bnlearn)
> data(asia)
> HITON.PC.E = learn.nbr(asia, 'E', method='si.hiton.pc',
                         alpha=0.01)
> HITON.PC.E
[1] "L" "T"
```

The result shows that L and T are in the parents and children set of the target variable E, which is consistent with the result from the global network in Fig. 3.2.

In the *bnlearn* package, mutual information test is set as the default conditional independence test for binary variables . However, we can specify a different type of conditional independence test for HITON-PC, e.g. Chi-square (denoted as "x2" in *bnlearn*) as follows:

```
> HITON.PC.E = learn.nbr(asia, 'E', method='si.hiton.pc',
                          test='x2', alpha=0.01)
> HITON.PC.E
[1] "L" "T"
```

3.4.2 Example 2: Using the Data Sets of Figs. 2.1 and 2.2

In this example, we firstly use the data set that has the same dependence relationships as the graph in Fig. 2.1. This data set can be downloaded from:

http://nugget.unisa.edu.au/Causalbook/

We now use HITON-PC to learn the local causal structure around node Z.

```
> library(bnlearn)
> data=read.csv("Example21.csv", header=TRUE, sep=",")
> data[1:5,]
  A B C D E F G H Z
1 1 1 1 0 0 1 1 1 1
2 1 1 1 0 0 1 0 1 1
3 1 0 1 1 0 1 1 1 1
4 1 0 0 0 0 0 0 1 0
5 1 0 1 0 0 1 1 1 1
> for(i in 1:9){data[,i] = as.factor(data[,i]) }
#bnlearn requires numeric or factor data types.
#the above command converts data of the nine
variables(nine columns)
#in the data set to factor data types.

> hiton.pc.Z=learn.nbr(data, 'Z', method='si.hiton.pc',
                        alpha=0.01)
> hiton.pc.Z
[1] "B" "F" "C"
```

The parents and children set of Z is $\{B, F, C\}$, which is consistent with the result discussed in Example 3.1.

We now use the data set that has the same dependence relationships as the Bayesian network shown in Fig. 2.2. The data set can be downloaded from:

http://nugget.unisa.edu.au/Causalbook/

This is a scenario when the HITON-PC algorithm will have false discoveries. We now use HITON-PC to learn the local causal structure around the target variable Z.

```
> library(bnlearn)
> data=read.csv("Example22.csv", header=TRUE, sep=",")
> for(i in 1:9){data[,i] = as.factor(data[,i])}
> hiton.pc.Z=learn.nbr(data, 'Z', method='si.hiton.pc',
                        alpha=0.01)
> hiton.pc.Z
[1] "C" "F" "B" "G"
```

Similar to the PC-simple algorithm, HITON-PC discovers G as a member of the parents and children set of the target Z, and this is a false discovery. On the other hand, the parents and children set of G is:

```
> hiton.pc.G=learn.nbr(data, 'G', method='si.hiton.pc',
                        alpha=0.01)
> hiton.pc.G
[1] "F" "E"
```

The above result shows that Z is not included in the parents and children set of G, therefore G will be removed from the parents and children set of Z when symmetry correction is used, and this eliminates the false discovery.

References

1. C. F. Aliferis, A. Statnikov, I. Tsamardinos, S. Mani, and X. D. Koutsoukos. Local causal and Markov blanket induction for causal discovery and feature selection for classification Part I: Algorithms and empirical evaluation. *Journal of Machine Learning Research*, 11:171–234, 2010.
2. C. F. Aliferis, A. Statnikov, I. Tsamardinos, S. Mani, and X. D. Koutsoukos. Local causal and Markov blanket induction for causal discovery and feature selection for classification Part II: Analysis and extensions. *Journal of Machine Learning Research*, 11:235–284, 2010.
3. M. Scutari. Learning Bayesian networks with the bnlearn R package. *arXiv preprint arXiv:0908.3817*, 2009.
4. I. Tsamardinos, L.E. Brown, and C.F. Aliferis. The max-min hill-climbing Bayesian network structure learning algorithm. *Machine Learning*, 65(1):31–78, 2006.

Chapter 4
Causal Rule Discovery with Partial Association Test

Abstract Partial association test is a statistical means to test whether the association between two variables is persistent given other variables. The CR-PA algorithm has been developed to identify causal rules (CRs) by integrating association rule mining and partial association (PA) tests. The use of association rule mining enables fast identification of causal hypotheses (association rules) from large data sets, and partial association tests on these association rules eliminate non-persistent associations. This chapter firstly describes the basics of partial association tests and association rule mining, and then the CR-PA algorithm is presented in detail, followed by the discussions on the complexity and false discoveries of the algorithm. A tool which implements CR-PA is also introduced.

4.1 Introduction

PC-simple and HITON-PC are developed on the ground of Bayesian network learning theory, and both algorithms use conditional independence tests to eliminate non-persistent associations.

Partial association [3, 9] is a statistical approach to testing conditional independence, and its idea is rooted in controlled experiments. To test the relationship between two variables, X and Z, while controlling the effects of a set of covariates, we should choose data samples that have similar distribution on the values of control variables or covariates (which is a subset of all variables other than X and Z) for the study. Partial association tests are often used to assess if an association between two variables X and Z found using the whole data set actually consistently exists across all the strata or sub-populations.

Often a partial table is used to represent the distribution of X and Z in each sub-population. For a value or level of the control variables, for example (Education=*college*, Occupation=*manager*), the partial table looks like the one shown below, where n_{ijk} $(i, j \in \{1, 2\})$ is the count of 'Gender=i' and 'Salary=j' at the kth value of the control variables, e.g. {*college, manager*}.

© The Author(s) 2015
J. Li et al., *Practical Approaches to Causal Relationship Exploration*,
SpringerBriefs in Electrical and Computer Engineering, DOI 10.1007/978-3-319-14433-7_4

A partial association test then uses the partial tables at all values of the control variables to obtain a test statistic to assess the partial association between X and Z over all sub-populations. Within each sub-population, as the control variables have the same values in all samples, their effect (if any) on the association between X and Z is removed. Therefore if there exists significant partial association between X and Z in all sub-populations, it is likely that the association between X and Y is genuine and persistent. This means that it is reasonable to apply partial association tests to evaluate the persistence of associations for causal discovery.

Let us complete the previous example. We try to determine whether gender difference causes salary difference. Let us assume that all other relevant variables are Education and Occupation, which have only two values each, i.e. {*college*, *high school*} for Education and {*manager*, *clerk*} for Occupation. with partial association test, we will need to evaluate the association between Gender and Salary in four partial tables which summarise the statistics of the four sub-populations: managers with college education, clerks with college education, managers with high school certificates, and clerks with high school certificates. If the association is persistent in all four sub-populations, we conclude that Gender is a cause of Salary.

To use partial association tests for causal discovery around a given target, we need firstly have a hypothesised cause variable and know the corresponding control variables. However, when there is no such knowledge for a given data set, especially when the number of variables in the data set is big, it is computationally inefficient to test the partial association between each predictor variable and the target.

The CR-PA (Causal Rule-Partial Association) algorithm adopts association rule mining to generate hypothesised causal relationships around a given target (represented as association rules). Then the algorithm conducts partial association test on each identified association rule, and considers the rule that represents a significant partial association as a causal rule.

In the following, before presenting the CR-PA algorithm, some background knowledge, including partial association tests and association rule mining are introduced.

4.2 Partial Association Test

The Mantel–Haenszel test [9] is commonly used for testing the null hypothesis of zero partial association between two variables X and Z in any of the strata of a population, against the alternative that the degree or strength of partial association is greater than 0. For the partial table shown in Table 4.1, the degree of partial association is defined as $\ln n_{11k} + \ln n_{22k} - \ln n_{12k} - \ln n_{21k}$ [3], i.e. the logrithm of the odds ratio for the partial table. In [3] Birch showed that the Mantel–Haenszel test was optimal when it was used to test an alternative hypothesis that the association in all strata is persistent.

The Mantel–Haenszel test statistic (with continuity correction) is defined as follows.

Table 4.1: An example partial table given $c_k = (Education = college,$ Occupation $= manager)$

$c_k = (college, manager)$	Salary=$high$	Salary=low	Total
Gender=$female$	n_{11k}	n_{12k}	$n_{1.k}$
Gender=$male$	n_{21k}	n_{22k}	$n_{2.k}$
Total	$n_{.1k}$	$n_{.2k}$	$n_{..k}$

Definition 4.1 (The Mantel–Haenszel test statistic). For a binary predictor variable X, a binary target variable Z, and a set of control variables $\mathbf{C}$ with r levels (distinct values), let T_k $(1 \leq k \leq r)$ be the partial table of X and Z given $\mathbf{C}$ at level k, where n_{11k}, n_{12k}, n_{21k} and n_{22k} are observed frequencies of X and Z, then the Mantel–Haenszel test statistic is:

$$MH = \left(| \sum_{k=1}^{r} \frac{n_{11k}n_{22k} - n_{21k}n_{12k}}{n_{..k}} | - \frac{1}{2} \right)^2 / \sum_{k=1}^{r} \frac{n_{1.k}n_{2.k}n_{.1k}n_{.2k}}{n_{..k}^2 (n_{..k} - 1)} \quad (4.1)$$

It has been shown that the Mantel–Haenszel test statistic has a Chi-square distribution (degree of freedom=1). Given a significance level α, if $MH \geq \chi_{\alpha}^2$, we reject the null hypothesis of independence, and consider that partial association between X and Z is significant. With the commonly used significance level $\alpha = 0.05$, $\chi_{\alpha}^2 = 3.84$.

Table 4.2: **a** An example data set for illustrating the Mantel–Haenszel test; **b** Summary of the example data set

X B D E	Z	count
1 0 0 1	0	5
0 0 0 1	0	8
0 0 0 1	1	4
1 0 0 1	1	15
0 0 1 0	1	15
0 0 1 1	0	3
0 1 1 0	0	8
1 1 1 0	1	14
0 1 1 0	1	3
1 1 1 0	0	8

(a)

X B D E	Z = 1	Z = 0
0 0 0 1	4	8
0 0 1 0	15	0
0 0 1 1	0	3
0 1 1 0	3	8
1 0 0 1	15	5
1 1 1 0	14	8

(b)

With the example data set shown in Table 4.2, let us use the Mantel–Haenszel test to assess the partial association between X and Z. Control variable set $\mathbf{C} = \{B, D, E\}$, and it has four values: $c_1 = (B = 0, D = 0, E = 1)$, $c_2 = (B = 0, D = 1, E = 0)$, $c_3 = (B = 0, D = 1, E = 1)$ and $c_4 = (B = 1, D = 1, E = 0)$. Correspondingly there are four partial tables for X and Z. The two partial tables given c_1 and c_4 are listed below. Note that the other two partial tables (given c_2 and c_3) each has one row of all zero counts, and they do not contribute to the test statistic and thus are omitted.

$(B=0, D=0, E=1)$	$Z=1$	$Z=0$
$X=1$	15	5
$X=0$	4	8

$(B=1, D=1, E=0)$	$Z=1$	$Z=0$
$X=1$	14	8
$X=0$	3	8

For the partial table for $c_1 = (B=0, D=0, E=1)$, we have,

$$\frac{n_{11k}n_{22k} - n_{21k}n_{12k}}{n_{..k}} = \frac{15 \times 8 - 4 \times 5}{15 + 5 + 4 + 8} = 3.13$$

$$\frac{n_{1.k}n_{.1k}n_{.2k}n_{2.k}}{n_{..k}^2(n_{..k}-1)} = 1.87$$

For the partial table for $c_4 = (B=1, D=1, E=0)$, we have,

$$\frac{n_{11k}n_{22k} - n_{21k}n_{12k}}{n_{..k}} = \frac{14 \times 8 - 3 \times 8}{14 + 8 + 3 + 8} = 2.67$$

$$\frac{n_{1.k}n_{.1k}n_{.2k}n_{2.k}}{n_{..k}^2(n_{..k}-1)} = 1.89$$

According to Formula (4.1), $MH = (3.13 + 2.67 - 1/2)^2/(1.87 + 1.89) = 7.47$. Suppose that $\alpha = 0.05$, as the test statistic is greater than 3.84, the partial association between X and Z is significant.

In this example, as the control variable set $\mathbf{C}$ contains 3 variables, there are $2^3 = 8$ possible values or levels of the variables, thus there are up to eight 2×2 partial tables. However, we have seen that the number of the levels which have records in the data set can be smaller. In this example only 4 out of the 8 possible levels appear in Table 4.2. Furthermore a partial table that has all zeros in one row or one column can be eliminated because the contribution to the test statistic from such a table is zero. As a result, the actual number of valid partial tables could be much smaller. With this example, only two partial tables are actually used in the calculation of the test statistic.

4.3 Association Rule Mining

Association rule mining [1] is a popular data mining approach to uncovering interesting relationships between values of variables. An association rule, in the form of $(X = 1) \to (Z = 1)$, represents an association between $X = 1$ and $Z = 1$. As a well studied area in data mining, many efficient association rule mining algorithms have been developed [5]. Therefore for causal discovery in large data sets, it is practical to use association rule mining in the preliminary investigations to build causal hypotheses (candidate causal relationships). Another advantage of using association rule mining in causal discovery is that the left hand side of an association rule may be a combination of multiple variable values, for example, in the rule $\{male, 40 - 50\} \rightarrow depression$, the left hand side contains two values of gender and age range respectively. Finding such association rules can facilitate the identification of combined causes in which multiple factors work together to lead to an effect.

In the following, we firstly describe the notation and concepts related to association rule mining.

Let $\mathbf{X} = \{X_1, X_2, \ldots, X_m\}$ be a set of (binary) predictor variables and Z be the target variable. $(x_1, x_2, \ldots, x_m)$ represents that $(X_1 = 1, X_2 = 1, \ldots, X_m = 1)$ and z represents $Z = 1$. $x_1, x_2, \ldots, x_m$ and z are known as *items*. In association rule notation, value 1 of a variable indicates the presence of an event and value 0 of a variable indicates the absence of an event. When we are interested in the presence of events, an item is equivalent to a binary variable. Therefore, we use item x and variable X interchangeably. Binary variables can represent variables of multiple discrete values. For example, a multi-valued data set for the variables (Gender, Age, ...) is equivalent to a binary data set for the variables (Male, Female, 0-19, 20-39, 40-69, ...). Both Male and Female variables are kept to allow us to have combined variables that involve them separately, for example, (Female=1, 40-59=1, Diabetes=1) and (Male=1, 40-59=1, Smoking=1).

We call X a *combined* variable if it is a combination of multiple variables $X_1, \ldots, X_l$ $(l \geq 2)$ such that $X = 1$ if and only if $(X_1 = 1, \ldots, X_l = 1)$, and $X = 0$ otherwise. Alternatively x is an *itemset* containing multiple items, i.e. $x = x_1 \ldots x_l$.

Definition 4.2 (Support of itemset and frequent itemset). Given a data set, the *support* of an itemset x is the number of records in the data set that contains x, denoted as $supp(x)$. With a specified minimum support threshold, m_{supp}, if $supp(x) > m_{supp}$, x is a *frequent itemset*, otherwise x is an *infrequent itemset*.

Definition 4.3 (Support and confidence of a rule). A *rule* is in the form of $(X = 1) \rightarrow (Z = 1)$, or $x \rightarrow z$, and let LHS$(x \rightarrow z) = x$ and RHS$(x \rightarrow z) = z$. The support of rule $x \rightarrow z$ is the number of records in a data set containing both x and z, denoted as $supp(x \rightarrow z)$. $supp(x \rightarrow z) = supp(xz)$. The *confidence* of a rule is defined as $supp(x \rightarrow z)/supp(x)$.

Traditionally an association rule is defined as a rule that meets the minimum requirement of support and confidence [1]. However, it has been shown that such an association rule may not indicate a real association between the LHS and RHS of a rule [4].

CR-PA chooses to use Chi-square test of independence for determining the existence of associations between two variables, and defines an association rule as follows.

Definition 4.4 (Association rule). Give a data set D, let m_{supp} be the minimum support threshold and α be the significance level of Chi-square independence test, $x \rightarrow z$ is an association rule if $supp(x \rightarrow z) > m_{supp}$, i.e. xz is a frequent itemset, and X and Z are associated, i.e. $\chi_D^2 \geq \chi_\alpha^2$, where χ_D^2 is the Chi-square test statistic and χ_α^2 is the critical value of the Chi-square distribution for α.

As the LHS of an association rule can be a combined variable, the search space for association rules is exponential to the number of variables included in a given data set. Fortunately, many techniques for pruning the search space have been developed. The corner stone of these pruning techniques is the anti-monotone property of rules. A commonly used anti-monotone property is given below.

Property 4.1. If x is an infrequent itemset, then any super itemset of x (i.e. any superset of x) will be infrequent too.

This property implies that once we know that x is an infrequent itemset, the search space can be pruned by ignoring all the itemsets that are supersets of x.

Another way of improving the efficiency of association rule mining is to use proper data structure in algorithm implementations to facilitate fast search and counting. For this purpose, CR-PA employs a prefix tree in its implementation. Interesting readers can refer to [8] for the details of using prefix trees in association rule mining.

4.4 The CR-PA Algorithm

In this section, we firstly define causal rules using the concepts of association rules and partial association.

Definition 4.5 (Causal rule). Association rule $x \to z$ is a causal rule if there exists a significant partial association between variables X and Z.

Note that in this definition x is an itemset, i.e. X can be a combined variable. As described above, the anti-monotone property (Property 4.1) can be applied to prune the search space. Because our final goal is to discover causal relationships, we can further prune the search space based on the following observations of causal relationships.

Observation 4.1 *If $x \to z$ is a causal rule, then any more specific causal rules whose LHS contains x, e.g. $xp \to z$, is redundant.*

We see that the addition of more specific information does not impact the occurrence of the effect. For example, if *college graduate $\to$ high salary* is a causal rule, then it is clear that both male college graduates and female college graduates have high salaries, thus it is redundant to have the more specific rules *male college graduate $\to$ high salary* or *female college graduate $\to$ high salary*.

Observation 4.2 *If $x \to z$ is an association rule but it is not a causal rule, $xp \to z$ should not be tested for a causal rule.*

In the situation described in Observation 4.2, the association between x and z is mediated by other variables or both x and z are caused by another variable. In other words, there is no direct association between x and z. It is unlikely that x combining with another variable p will lead to a direct association.

The above two observations lead to the following pruning criterion.

Criterion 4.1 *It is possible for xy → z to be a causal rule if X and Z are not associated, and Y and Z are not associated either.*

For example, Drug 1 does not lead to an adverse drug reaction, and neither does Drug 2, but the use of both Drug 1 and Drug 2 may cause an adverse drug reaction.

CR-PA uses this criterion to forward prune the search space for causal rules such that all variables associated with the target variable are excluded from being considered as components of combined causes in future search.

The outline of CR-PA is shown in Algorithm 4.1. The algorithm produces causal rules level by level in terms of the number of items in the LHS of a rule. That is, in each iteration of the **while** loop (Lines 2–17), a set of causal rules with k (≥ 1) items in the LHS are detected and added to the output set **R**.

The candidate cause variables are stored in set **V**, which initially contains all the (single) predictor variables (Line 1). To obtain Level 1 causal rules, CR-PA firstly detects all the Level 1 association rules (Lines 3–10).

The *Frequent*() function (Line 3) removes a variable X from **V** if itemset xz is infrequent. This pruning reduces the number of variables in the tests of association or partial association with Z. More importantly, based on Property 4.1, all supersets of an infrequent itemset are infrequent, in the later step (Line 16) for generating next level candidate rules, the removed variables are not considered, which can significantly reduce of the number of candidate rules overall.

Each frequent itemset in **V** undergoes the Chi-square test (the *Association(X, Z)* function in Line 5). X is put into **P** if it is associated with Z (Line 6); otherwise it is placed in **N** (Line 8).

For each variable X in **P**, the function *PAassociation*() (Line 12) tests the partial association between X and Z. If X is partially associated with Z, X is added to **R** (Line 13), and $x \to z$ is a causal rule.

At the end of Level 1 iteration (Line 16), following Criterion 4.1, the *Generate*() function takes all the variables in **N**, i.e. variables that failed the association tests, to form Level 2 candidate rules, each containing two items in its LHS. For higher level candidate generation (higher than 2), Apriori candidate generation [2] is used for efficient candidate generation and pruning.

Next the variable sets **P** and **N** are reset to empty sets and Level 2 iteration of the **while** loop starts.

The iterations continue until no more higher level combined rules can be generated or the maximum length of rules $l_{\max}$ is reached. Finally **R** contains all identified causal rules.

Example 4.1. In the following we use the example data set in Table 4.3 to illustrate the main steps of CR-PA. The data set contains 961 records for three predictor variables A, B, C and the target variable Z. It was generated using a Bayesian network with the causal structure $B \to A \to Z$, and C is independent of A, B and Z. Assume that the minimum support threshold is 48, approximately 5 % of the total number of records, and the significance level for both Chi-square independence test and the Mantel–Haenszel test is 0.05.

Algorithm 4.1: Causal Rule discovery with Partial Association test (CR-PA)
[6]

Input: **D**, a data set for the set of predictor variables $\mathbf{X} = \{X_1, X_2, \ldots, X_m\}$ and the target
variable Z; l_{max} the maximum number of values in the LHS of a rule; m_{supp} the minimum
support; and α, significance level for conditional independence test and the
Mantel–Haenszel test.

Output: **R**, a set of cause rules of which the LHS are Z

 1: let **P**$=\emptyset$, **N** $= \emptyset$, **R** $= \emptyset$, $k = 1$, **V** $= \mathbf{X}$, where **P** stores itemsets associated with Z, **N** stores
      itemsets that are not associated with Z, and **V** stores LHS of candidate rules.
 2: **while** (**V** is not empty AND $k \leq l_{max}$) **do**
 3:        **V** $\leftarrow$ Frequent(**V**)
 4:        **for** each $X \in \mathbf{V}$ **do**
 5:              **if** Association(X, Z) is true **then**
 6:                    insert X to **P**
 7:              **else**
 8:                    insert X to **N**
 9:              **end if**
10:        **end for**
11:        **for** each $X \in \mathbf{P}$ **do**
12:              **if** PAssociation(X, Z) is true **then**
13:                    insert X to **R**
14:              **end if**
15:        **end for**
16:        **V** $\leftarrow$ Generate(**N**), and let $k = k + 1$, **P** $= \emptyset$, **N** $= \emptyset$
17: **end while**
18: output **R**

Table 4.3: **a** The data set for illustrating CR-PA; **b** The itemised view of the data set

A B C Z	count
0 0 0 0	61
0 0 0 1	18
0 0 1 0	168
0 0 1 1	28
0 1 0 0	39
0 1 1 0	100
0 1 1 1	24
1 0 0 0	15
1 0 0 1	47
1 0 1 0	21
1 0 1 1	111
1 1 0 1	81
1 1 1 0	57
1 1 1 1	191

(a)

item	count
	61
z	18
c	168
$c\ z$	28
b	39
$b\ c$	100
$b\ c\ z$	24
a	15
$a\ \ \ \ z$	47
$a\ \ c$	21
$a\ \ c\ z$	111
$a\ b\ \ z$	81
$a\ b\ c$	57
$a\ b\ c\ z$	191

(b)

Firstly we call the *Frequent*() function. From Table 4.3, we have $supp(az) = 430$, $supp(bz) = 296$, and $supp(cz) = 354$. As all the supports are greater than the given minimum support, a, b, and c are all frequent items and the three predictor variables are kept in **V** for the Chi-square independence tests.

For the three rules $a \rightarrow z$, $b \rightarrow z$ and $c \rightarrow z$, from Table 4.3, the contingency tables are ($\bar{x}$ indicates item x does not occur in a record):

	z	$\bar{z}$	total
a	430	93	523
$\bar{a}$	70	368	438
total	500	461	961

	z	$\bar{z}$	total
b	296	196	492
$\bar{b}$	204	265	469
total	500	461	961

	z	$\bar{z}$	total
c	354	346	700
$\bar{c}$	146	115	261
total	500	461	961

Then the Chi-square test statistics (χ^2) are 419.0, 26.7, and 2.2 respectively. So the first two rules are accepted as association rules as their test statistics are greater than the critical value, 3.84, and c is included in **N**.

In the next step, CR-PA tests if the two association rules are causal rules, i.e. if a and b are partially associated with z.

To test if $a \rightarrow z$ is a causal rule, four partial tables are formed corresponding to the four itemsets (four levels of control variables, B and C), $\emptyset$, b, c and bc:

$\emptyset$	z	$\bar{z}$
a	47	15
$\bar{a}$	18	61

b	z	$\bar{z}$
a	81	0
$\bar{a}$	0	39

c	z	$\bar{z}$
a	111	21
$\bar{a}$	28	168

bc	z	$\bar{z}$
a	191	57
$\bar{a}$	24	100

According to Formula (4.1) the test statistic $MH = 400.2$, which is greater than the critical value. This indicates that $a \rightarrow z$ is a causal rule.

Similarly, for the test of rule $b \rightarrow z$, we have the following four partial tables for the four itemsets, $\emptyset$, a, c and ac:

$\emptyset$	z	$\bar{z}$
b	0	39
$\bar{b}$	18	61

a	z	$\bar{z}$
b	81	0
$\bar{b}$	47	15

c	z	$\bar{z}$
b	24	100
$\bar{b}$	28	168

ac	z	$\bar{z}$
b	191	57
$\bar{b}$	111	21

According to Formula (4.1) the test statistic $MH = 0.001$, which is lower than the critical value. This indicates that $b \rightarrow z$ is not a causal rule.

Since only c is in **N**, there is no Level 2 candidate rule generated, so the **while** loop stops and one causal rule $a \rightarrow z$ is identified at the end.

This result is consistent with the model used to generate the data set. The association between $b \rightarrow z$ is intermediated by a and hence is not persistent and does not indicate causality.

4.5 Discussions

Association rules show relationships between values whereas Bayesian based methods find relationships between variables. We can consider a value itself as a binary variable where 1 stands for the presence and 0 for the absence. So, we still use variables in the following discussions to be consistent with the discussions in previous chapters.

4.5.1 Complexity of CR-PA

The time complexity of CR-PA is due to the two major tasks: association rule mining and partial association tests.

Given a data set of m predictor variables and the target variable, the search space of association rules is $\sum_{i=1}^{l_{max}} \binom{m}{i}$, where l_{max} is the maximum number of items in the LHS of a rule. When $l_{max} \ll m$, the time complexity is approximately $O(m^{l_{max}})$. For one partial association test, in the worst case when all other predictor variables are control variables, the number of partial tables is 2^{m-1}. Therefore, the time complexity of a naive algorithm is exponential to the number of variables. Therefore only for a low dimensional data set with a very small number of variables, e.g. the one in Example 4.1, we may use all other predictor variables as control variables.

However, as we have seen from the previous section, CR-PA prunes the search space by removing non-frequent itemsets. Furthermore, the generation of higher level candidate rules only considers variables that are not associated with the target. These measures in practice reduces the complexity significantly.

Additionally, with one partial association test, e.g. between X and Z given control variables $\mathbf{C}$, for each level of $\mathbf{C}$, instead of sifting through the data set to get the partial tables (in the worst case $\mathbf{C}$ has $O(2^m)$ levels), CR-PA sorts the data set based on the values of $\mathbf{C}$ using quick sort, which has time complexity of $O(n \log n)$, where n is the number of records in the data set. In the sorted data set, records with same values for control variables are listed contiguously, so the algorithm only needs to pass the sorted data set once to get all the partial tables for calculating the test statistic, and the time complexity for a partial association test becomes $O(n \log n)$.

Therefore by using effective pruning for causal rule mining and quick sort for finding partial tables, the efficiency of CR-PA is high.

When we are only interested in causal rules with single item in their LHS, CR-PA in fact is very efficient. In this case, the number of association rules is at most m, and based on the above analysis of the time taken for a partial association test, the time complexity of CR-PA is $O(mn \log n)$.

4.5.2 False Discoveries of CR-PA

We discuss the false discoveries of CR-PA related to algorithm design and input data.

The major source of false discoveries is the selection of control variables. We will present more discussions on control variables in the following chapter. In this chapter, control variables can be simply considered as variables to partition a data set into sub-populations and form partial tables for partial association tests.

For low dimensional data sets (with a few variables), we may not need to select the control variables. However, in order to have 5 samples in each sub-population for a partial association test, with n control variables, 5×2^n data records are needed. For example, when 2, 4, 10, 20, and 30 control variables are used, the number of

Table 4.4: Examples showing that variables strongly associated with LHS of a rule make partial association test impossible. The rule to be tested is $a \to z$. **a** D and A are negatively associated; **b** E and A are positively associated

A D B C Z	count
0 1 0 0 0	61
0 1 0 1 0	168
0 1 0 1 1	28
0 1 1 0 0	39
0 1 1 1 0	100
0 1 1 1 1	24
1 0 0 0 0	15
1 0 0 0 1	47
1 0 0 1 0	21
1 0 0 1 1	111
1 0 1 1 0	57
1 0 1 1 1	15

(a)

A E B C Z	count
0 0 0 0 0	61
0 0 0 0 1	18
0 0 0 1 0	168
0 0 0 1 1	28
0 0 1 0 0	39
0 0 1 1 0	100
0 0 1 1 1	24
1 1 0 0 0	15
1 1 0 0 1	47
1 1 0 1 0	21
1 1 0 1 1	111
1 1 1 1 0	57
1 1 1 1 1	15

(b)

records required are 20, 80, 5120, 5242880, and 5368709120 respectively. So data insufficiency becomes a major problem with a slight increase of the number of variables, and with most real world data sets, selecting a set of control variables is essential for CR-PA to work. In the implementation of CR-PA, a variable is chosen to be a control variable if it is associated with the target variable (called a relevant variable). When the set of relevant variables is large, we are only able to include a limited number of most relevant variables in the control variable set to avoid the problem caused by data insufficiency. Such a choice may cause false positives.

Using variables that are highly positively or negatively associated with the LHS of a rule as control variables will cause false negatives. For example, with the two data sets in Table 4.4, D and E are positively and negatively associated with A respectively. When we test rule $a \to z$, all partial tables formed from the data set will have one row of zero counts (see below for two examples), hence the partial association test statistic will be zero.

$(D = 1, B = 1, C = 1)$	$Z = 1$	$Z = 0$
$A = 1$	0	0
$A = 0$	24	100

$(E = 1, B = 1, C = 1)$	$Z = 1$	$Z = 0$
$A = 1$	15	57
$A = 0$	0	0

In general, incorrect selection of control variables will cause false discoveries as discussed in Sect. 5.5.2, and we refer readers to that section for details.

Similar to PC-simple and HITON-PC, CR-PA assumes causal sufficiency, i.e. no common causes of two observed variables are not observed. This indicates that if the assumption does not hold for a given data set, the algorithm will have false discoveries.

The reliability of CR-PA depends on the input data set, because the statistical tests require complete or randomly sampled data, as well as sufficient number of samples to obtain correct test results.

An incorrect partial association test only affects the rule being tested itself such that the rule is incorrectly accepted as a causal rule (false positive) or mistakenly discarded (false negative). In theory, the Mantel–Haenszel test has a favourable property that it is not sensitive to small counts in partial tables [7]. However, partial tables with one row (or one column) of zero counts do not contribute to a test result and they will cause false discoveries.

When an association test is wrong, it will affect the results of the next level, as CR-PA combines the variables that are not associated with the target when generating the next level candidate rules. This is mainly due to the use of Observation 4.2. The Observation makes CR-PA efficient but sensitive to the correctness of association test when finding combined causes. The algorithm presented in the following chapter does not use the Observation, so the algorithm is slower than CR-PA, but it is not so sensitive to the threshold of association tests when finding combined causes.

4.6 An Implementation of CR-PA

A Java implementation of CR-PA is provided as part of the software tool, Causal Relationship Explorer (CRE), which can be downloaded from:

http://nugget.unisa.edu.au/Causalbook/

In the following we use examples to illustrate the steps involved in running CR-PA in the software tool.

4.6.1 Example 1: Running CR-PA in CRE

In this section, we use the data set discussed in Example 4.1, which can be downloaded from:

http://nugget.unisa.edu.au/Causalbook/

The input files are in C4.5 format, including *Example41.data* and *Example41.names*. With this data set, we can follow the steps below to run CR-PA.

1. Start CRE by double-clicking on the *CRE.jar* file. Note that Java SE runtime environment (jre7 or later) needs to be installed on the computer before running CRE. The main user interface of CRE is shown in Fig. 4.1.
2. Open the input file. Select on the menu bar, File → Open File. Select the input *.names* file stored on local computer, i.e. *Example41.names* in this case.
3. Select the algorithm CR-PA from the left panel.
4. Set parameters for CR-PA. On the menu bar, select Configuration → Parameters. A pop-up window will appear for setting CR-PA parameters as shown in Fig. 4.2. There are four parameters as follows:

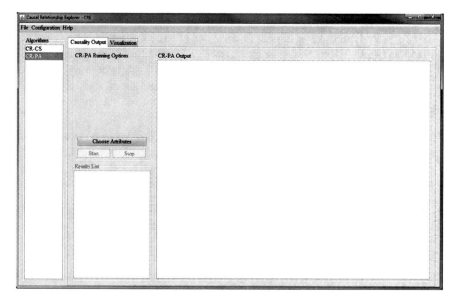

Fig. 4.1: The main user interface of CRE

- *Max level of combined rules*: l_{max} in Algorithm 4.1, the maximum level of combined causes, e.g. select 2 for mining single and combined (level 2) causal rules.
- *MinSupport*: the minimum support, i.e. m_{supp} in Algorithm 4.1.
- *Chi-square confidence level*: the confidence level of the Chi-square tests, i.e. $1 - \alpha$, with α in Algorithm 4.1.
- *PA confidence level*: the confidence level for the partial association tests. By default, this value is set to the same as the Chi-square confidence level.

Click "Confirm" after all of the parameters are set.

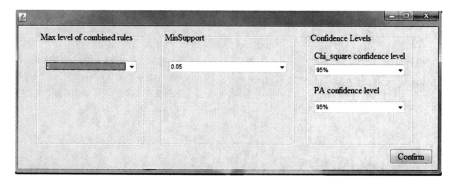

Fig. 4.2: The interface for parameter setting in CR-PA

5. Select attributes from the data set. As shown in Fig. 4.3, click the "Choose Attributes" button in the middle panel of the CRE user interface to select the attributes (predictor variables) of interest, then click the "All" button to select all attributes in the data set, or tick the boxes to choose a subset of attributes. Note that the target variable, which is the last column in the data set, is not included in the list of attributes. Click "Confirm" after selecting the desired attributes.

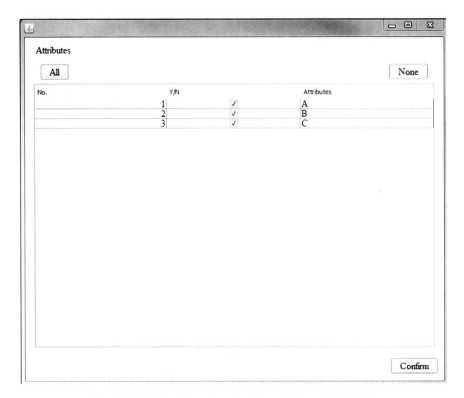

Fig. 4.3: The interface for choosing attributes in CR-PA

6. Run CR-PA. Click the "Start" button to run the CR-PA algorithm.
7. Obtain the results. The results are shown in the CR-PA output panel as in Fig. 4.4. Moreover, the results are also stored in the file *current_optforUI.csv*, located in the current folder of *CRE*. Each row of the report is a rule. The "Causal/Noncausal" column tells us if the rule is causal rule or just an association rule. The "level" column shows the length of the LHS of a rule, with level 1 indicating a single variable in the LHS of a rule and level 2 indicating combined (level 2) rules. The last four columns show the statistics of the rule which are the values of the four cells in the contingency table. We can see from the result that with the example data set there is only one causal rule at level one, $A = 1 \rightarrow Z = 1$, which is consistent with the causal structure $B \rightarrow A \rightarrow Z$

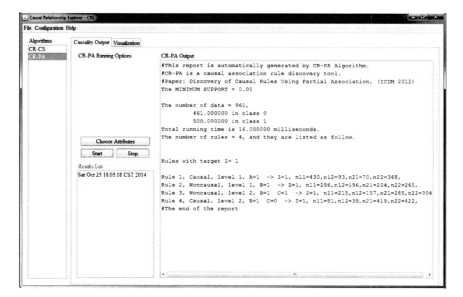

Fig. 4.4: The output of CR-PA algorithm for the data set in Example 4.1

in Example 4.1. Note that in this example, we are only interested in rules with $Z = 1$. To generate rules with $Z = 0$, we need to swap the positions of "0" and "1" at the first line of the *Example41.names* file before running CR-PA.

4.6.2 Example 2: Using the Data Sets of Figs. 2.1 and 2.2

The data set for the Bayesian network shown in Fig. 2.1 in Chap. 2 can be downloaded from: *http://nugget.unisa.edu.au/Causalbook/*. This data set is the same as the one used in Chaps. 2 and 3, but it has been converted into the C4.5 format, which includes the two files *Example21.data* and *Example21.names*. With this data set, PC-simple and HITON-PC have correctly identified B, C, and F as the parents or children of Z. Now, we apply CR-PA to the data set to explore causal rules.

Using the same procedure as for the above example, we set the maximum level of combined variables to 2, and keep the default setting as shown in Fig. 4.2 for other parameters. The result is shown in Fig. 4.5.

We can see that B, C, and F are found to have causal relationships with the target Z, which is consistent with the results by PC-simple and HITON-PC. The relationships represented in rules are specific. For example, at the first level, we have $C = 1 \rightarrow Z = 1$, $B = 0 \rightarrow Z = 1$, and $F = 0 \rightarrow Z = 1$.

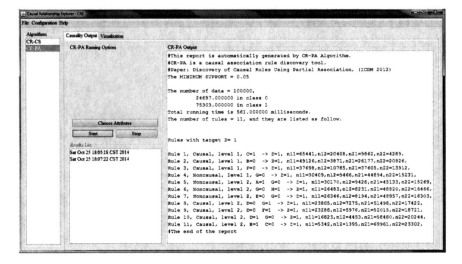

Fig. 4.5: The output of CR-PA for the data set in Fig. 2.1

Similarly, we apply CR-PA to the data set of Fig. 2.2, which can be downloaded from *http://nugget.unisa.edu.au/Causalbook/*. We use the same settings as above, and the result is shown in Fig. 4.6.

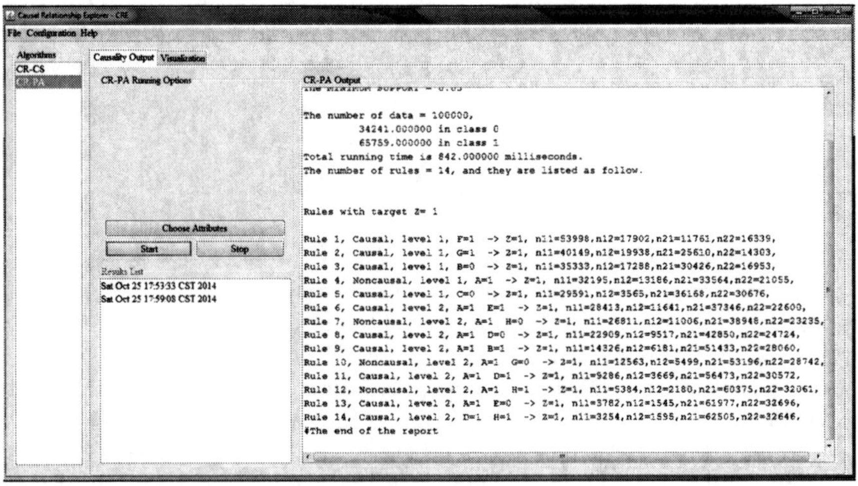

Fig. 4.6: The output of CR-PA for the data set in Fig. 2.2

With this data set, PC-simple and HITON-PC discovered the parents and children set of Z as $\{B,C,F,G\}$, which involves a false discovery, G, as discussed in Chapters 2 and 3. CR-PA also identifies four causal rules at level 1, $F = 1 \rightarrow Z = 1$, $G = 1 \rightarrow$

$Z = 1$, $B = 0 \rightarrow Z = 1$, and $C = 0 \rightarrow Z = 1$. Similar to PC-simple and HITON-PC, CR-PA has a false discovery $G = 1 \rightarrow Z = 1$. However, if we apply the symmetry correction as in Chapters 2 and 3 for CR-PA, we can remove the false discovery from the result. For example, we now set G as the target variables, and denote Z as Z' (to differentiate between the variable Z and the new target). The data set with the new target variable (G) is available at: *http://nugget.unisa.edu.au/Causalbook/*. We can see from the result (Fig. 4.7) that the variable Z (now Z') is not a cause of the target G. Therefore, based on the symmetry property, G will be removed from the list of causal rules with the target Z.

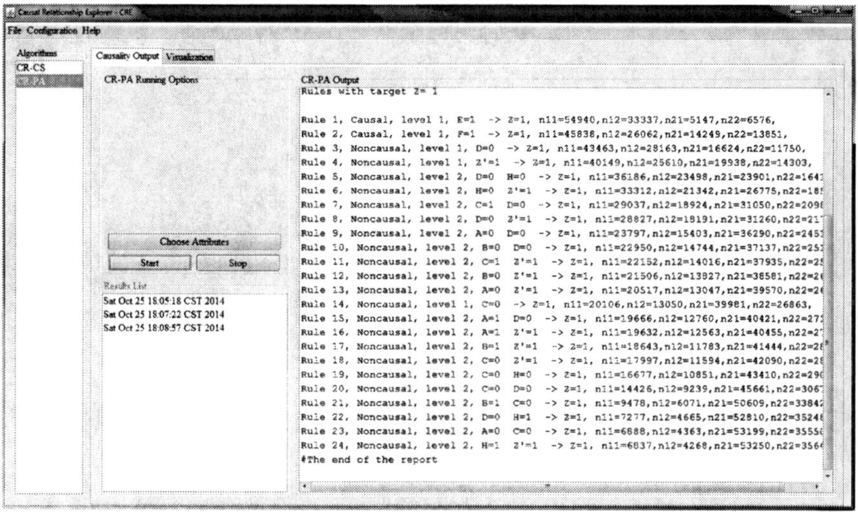

Fig. 4.7: The output of CR-PA for the data set in Fig. 2.2 with G as the target

References

1. R. Agrawal, T. Imieliński, and A. Swami. Mining association rules between sets of items in large databases. In *Proceedings of SIGMOD'93*, pages 207–216, 1993.
2. R. Agrawal, H. Mannila, R. Srikant, H. Toivonen, and A. I. Verkamo. Fast discovery of association rules. In *Advances in Knowledge Discovery and Data Mining*, volume 12, pages 307–328. AAAI/MIT Press Menlo Park, CA, 1996.
3. M. W. Birch. The detection of partial association, I: The 2×2 Case. *Journal of the Royal Statistical Society*, 26(2):313–324, 1964.
4. S. Brin, R. Motwani, and C. Silverstein. Beyond market baskets: Generalizing association rules to correlations. In *Proceedings of SIGMOD'97*, pages 265–276, 1997.
5. J. Han and M. Kamber. *Data Mining: Concepts and Techniques*. Morgan Kaufmann Publishers Inc., San Francisco, CA, USA, 2nd edition, 2005.

6. Z. Jin, J. Li, L. Liu, T. D. Le, B. Sun, and R. Wang. Discovery of causal rules using partial association. In *Data Mining (ICDM), 2012 IEEE 12th International Conference on*, pages 309–318, 2012.
7. S. J. Kuritz, J. R. Landis, and G. G. Koch. A general overview of Mantel-Haenszel methods: Applications and recent development. *Annual Review of Publish Health*, 9:123–160, 1988.
8. J. Li, J. Liu, H. Toivonen, K. Satou, Y. Sun, and B. Sun. Discovering statistically non-redundant subgroups. *Knowledge-Based Systems*, 67:315–327, 2014.
9. N. Mantel and W. Haenszel. Statistical aspects of the analysis of data from retrospective studies of disease. *Journal of the National Cancer Institute*, 22(4):719–748, 1959.

Chapter 5
Causal Rule Discovery with Cohort Studies

Abstract In this chapter, we present CR-CS, a different method for mining causal rules. Similar to CR-PA introduced in the previous chapter, CR-CS uses association rule mining to generate causal hypotheses, but for validating the hypotheses, CR-CS utilises the idea of a retrospective cohort study (CS). For each candidate causal rule, CR-CS firstly creates a fair data set which consists of pair-wise matched samples in the original data set and then it tests the association of the LHS (predictor) and the RHS (target) of the rule based on the fair data set. The creation of a fair data set imitates the process of collecting samples in a cohort study, for which each sample in the exposure group (LHS = 1) has a matching sample in the control group (LHS = 0) with respect to the values of control variables. Therefore an association confirmed using a fair data set is expected to be a persistent association across different values of the control variables, since matching has been employed to eliminate the effect of control variables. In this chapter, after introducing the motivation and general idea of CR-CS, the test of associations used by CR-CS is described. Then the CR-CS algorithm is presented in detail, followed by the discussions on the algorithm. Finally a tool which implements CR-CS is illustrated with examples.

5.1 Introduction

In order to test whether there exists a persistent association between a predictor variable X and the target variable Z, PC-simple and HITON-PC conduct conditional independence tests, essentially given each subset of all variables other than X and Z, i.e. they test the association between X and Z in all possible data segments created using each of the subsets of all other variables. Similarly partial association tests are based on data segments too, but the segmentation is done with only one set of variables, which is the set of control variables.

Testing associations across data segments allows us to examine if the association between X and Z is persistent with respect to different values of the variables other

J. Li et al., *Practical Approaches to Causal Relationship Exploration*,
SpringerBriefs in Electrical and Computer Engineering, DOI 10.1007/978-3-319-14433-7_5

than X and Z. In other words, such testing tries to remove the effect of other variables when validating a candidate causal relationship.

With observational data, another way of eliminating the effect of other variables is to use matching methods to create the control and exposure groups such that samples in the two groups have similar distributions over the values of control variables or covariates. This approach is often used in observational studies, including cohort studies and case-control studies [8].

The CR-CS method [6] adopts the idea of a retrospective cohort study when selecting samples and testing associations for causal discovery. In a retrospective cohort study, samples in the exposure group ($X = 1$) and the control group ($X = 0$) are selected from the observed data without knowing their responses. The key is that samples in the two groups need to be matched such that the distributions of the control variables in the two groups can be as close as possible. Then if we observe from the matched samples a strong association between the exposure variable X and the response or target variable Z, it is reasonable to claim that X is a cause of Z.

The CR-CS algorithm consists of three major steps. First, similar to CR-PA, association rule mining is employed to identify candidate causal rules. Then each candidate rule is assessed in the same way as in a retrospective cohort study. Specifically, a fair data set is created for the rule based on the given data set. In a fair data set, each sample in the exposure group has a matched sample in the control group in terms of the values of control variables according to certain similarity measure. Therefore fair data sets are similar to the data sets used in retrospective cohort studies in the sense that the distributions of control variables in the exposure and control groups are close to each other. In the final step, the association between the exposure variable and the target variable is tested using the fair data set. If the test result shows a strong association between the two variables, then the algorithm accepts the candidate rule as a causal rule.

In the following, we will firstly introduce the basic concepts involved in the main steps of CR-CS before presenting the algorithm in detail.

5.2 Association Rules Defined by Odds Ratio

Similar to CR-PA, the CR-CS method defines rule $x \rightarrow z$ as an association rule if its support is greater than the given minimum support and X and Z are associated.

CR-PA applies Chi-square test for independence to determine an association in order to be consistent with the Mantel-Haenszel test for determining partial association, which also uses a test statistic of Chi-square distribution. Because odds ratio has been widely used in retrospective studies [4] for validating a causal relationship, CR-CS chooses to use odds ratio for evaluating associations, and an association rule is defined as follows.

Definition 5.1 (Association rule (odds ratio based)). Given a data set **D**, let m_{supp} and m_{OR} be the minimum support and minimum odds ratio respectively, $x \rightarrow z$ is an association rule if $\text{supp}(x \to z) > m_{supp}$, and the odds ratio of the rule, $OR_\mathbf{D}(x \to z) > m_{OR}$.

The definition of the odds ratio of a rule is given below.

Definition 5.2 (Odds ratio of a rule). Given the following contingency table of a rule, $x \to z$,

the odds ratio of the rule $x \to z$ on $\mathbf{D}$ is defined as:

$$OR_\mathbf{D}(x \to z) = \frac{n_{11} * n_{22}}{n_{12} * n_{21}} \tag{5.1}$$

As the odds ratio of rule $x \to z$ is defined as the ratio of the odds of $Z = 1$ occurring in group $X = 1$ (i.e. n_{11}/n_{12}) and the odds of $Z = 1$ occurring in group $X = 0$ (i.e. n_{21}/n_{22}), an odds ratio that is equal to, greater than or less than 1 indicates that X and Z have a zero, positive or negative association respectively.

The above definition of association rules only concerns with positive associations. This inherits from traditional association rule definition. When we are interested in negative association rules, in odds ratio based association rules, we can keep rules whose odds ratio is less than $1/m_{OR}$ for the negative association rules.

Because statistically reliable associations do not always indicate causal relationships, CR-CS conducts a retrospective cohort study to validate candidate causal rules, for which the basic idea and definitions are presented in the next section.

5.3 Fair Data Sets and Causal Rules

Matching [9] is an important approach to causal inference in observational studies [2, 7]. To shield the effects of control variables when studying the causal relationship of two variables, namely exposure and target variables, researchers often choose two groups each taking a different values in the exposure variable while the values of the target variable are unknown or blinded. The distribution of the values of control variables in the two groups should be as close as possible. Cohort study [3, 8] is a typical example. In a cohort study, based on the status of being exposed to a potential cause factor (e.g. certain radiation), an exposure group of subjects and a non-exposure (or control) group of subjects are selected, and then followed to observe the occurrence of the outcome (e.g. cancer). In a retrospective cohort study, researchers look back at outcomes that have already recorded. To choose two groups with the same distribution in the control variables, a matching method is used to find identical (or very similar) records in terms of the values of control variables.

Constructing a fair data set is to simulate a cohort study in a data set for testing a hypothesis represented by an association rule. To define a fair data set, we firstly present the definition of a matched record pair.

Definition 5.3 (Matched record pair). Given an association rule $x \rightarrow z$ and a set of control variables C, a pair of records match if one contains $X = 1$, the other contains $X = 0$, and both have the matching values for C according to certain similarity measure.

The matching can be exact, in which a pair of records must have the same values for control variables. However, non-exact matching is also applicable and many similarity measures [5] can be used in matching pairs of records, e.g. Euclidean distance, Jaccard distance, Mahalanobis distance and propensity score [9], depending on the data in use and the specific problem considered.

Definition 5.4 (Fair data set for a rule). Given an association rule $x \rightarrow z$ that has been identified from a data set D and a set of control variables C, the fair data set D_f for the rule is the maximum sub data set of D that contains only matched record pairs from D.

The requirement of maximum sub data set of D in the definition enables maximal utilisation of the records in the original data set D.

The following example is used to illustrate the definition of fair data sets.

Example 5.1. Given an association rule $a \rightarrow z$ identified using the data set in Table 5.1a, and the control variable set $C = \{B, D, E, F\}$. Using exact matching, records (#1, #5), (#2, #6) and (#3, #8) form three matched pairs. A fair data set for $a \rightarrow z$ includes records (#1, #2, #3 #5, #6, #8) (see Table 5.1b). One record may have multiple matching records. For example, (#3, #7) and (#3, #8) both are matched pairs (in terms of record #3). So a fair data set for a rule may not be unique. When there are more than one matches for a record, CR-CS selects one at random. To avoid the variance in a result because of the random selection in creating a fair data set, it is better to run the CR-CS algorithm multiple times and output the causal rules with the majority votes in those runs.

After a fair data set is created for a rule, its odds ratio can be defined as follows [4].

Definition 5.5 (Odds ratio of a rule on a fair data set). Assume that D_f is a fair data set of association rule $x \rightarrow z$, then the odds ratio of the rule on D_f is:

$$OR_{D_f}(x \rightarrow z) = \frac{n_{12}^f}{n_{21}^f} \tag{5.2}$$

where n_{12}^f is the number of matched record pairs containing $Z = 1$ in the exposure group and $Z = 0$ in the control group, and n_{21}^f is the number of matched record pairs containing $Z = 0$ in the exposure group and $Z = 1$ in the control group. In the exposure group, $X = 1$ and in the control group, $X = 0$.

Table 5.1: **a** The original data set; **b** A fair data set for $a \rightarrow z$. The horizonal line separates the exposure group and the control group

ID	A	B	D	E	F	Z
1	1	0	1	0	0	1
2	1	0	1	0	1	1
3	1	1	0	1	0	0
4	1	1	0	0	0	1
5	0	0	1	0	0	0
6	0	0	1	0	1	0
7	0	1	0	1	0	0
8	0	1	0	1	0	1

(a)

ID	A	B	D	E	F	Z
1	1	0	1	0	0	1
2	1	0	1	0	1	1
3	1	1	0	1	0	0
5	0	0	1	0	0	0
6	0	0	1	0	1	0
8	0	1	0	1	0	1

(b)

For example, in Table 5.1b, $n_{12}^f = 2$ and $n_{21}^f = 1$ (note that we will need to use matched pairs to work out n_{12}^f and n_{21}^f). So $OR_{D_f}(a \rightarrow z) = 2$.

Based on the definition of the odds ratio over a fair data set, we can define a causal rule as follows.

Definition 5.6 (Causal rule). An association rule $(x \rightarrow z)$ indicates a causal relationship between X and Z and thus is called a causal rule, if its odds ratio on its fair data set, $OR_{D_f}(x \rightarrow z) > m_{OR}$, where m_{OR} is the given minimum odds ratio.

Based on Definition 5.6, testing whether an association rule is a causal rule becomes a problem of finding the fair data set for the rule. The search for a fair data set needs a control variable set.

The concept of control variable is the same as the control variable in a controlled experiment. When we study the relationship between a variable X and the target variable Z, another variable may affect the value of the target variable, thus it should be kept unchanged during the experiment. Such a variable is called a control variable.

A criterion for determining a control variable is that it should be correlated to the target variable. Since a correlated variable may be a cause of the target, hence it should be controlled. In addition, in terms of implementing CR-CS, inclusion of a variable that is highly associated (either positively or negatively) with the LHS of a rule in a control variable set results in a near empty fair data set (please see discussions in Sect. 5.5.2 for details). In the algorithm, both of the criteria have been considered.

To form a fair data set, we need the concept of equivalence classes.

Definition 5.7 (Equivalence class). Given a data set **D** and a set of control variables **C**, a maximum subset of the records of **D** which contain the same value of **C** form an equivalence class.

For example, in Table 5.1, records (#3, #7, #8) form an equivalence class given control variables B, D, E and F.

5.4 The CR-CS Algorithm

In this section we present the CR-CS algorithm (Algorithm 5.1) for causal rule mining.

Algorithm 5.1: Causal Rule mining with Cohort Study (CR-CS) [6]

Input: **D**, a data set for the set of predictor variables $\mathbf{X} = \{X_1, X_2, \ldots, X_m\}$ and the target variable Z; m_{supp}, minimum support; m_{OR}, minimum odds ratio; z', critical value used by significance test of odds ratio; l_{max}, the maximum length of a rule.

Output: **R**, a set of cause rules of which the RHS are Z.

 1: let causal rule set $\mathbf{R} = \emptyset$ and $k = 1$
 2: $\mathbf{X} \leftarrow \text{Frequent}(\mathbf{X})$
 3: find the set of irrelevant variables **I**
 4: let $\mathbf{X}_k = \mathbf{X}$
 5: **while** ($\mathbf{X}_k \neq \emptyset$ AND $k \leq l_{\text{max}}$) **do**
 6:     generate association rules $\{r_i\}$ at the k-th level, where $LHS(r_i) \in \mathbf{X}_k$
 7:     **for** each generated rule r_i **do**
 8:         find exclusive variables **E** of $LHS(r_i)$
 9:         let control variable set $\mathbf{C} = \mathbf{X} \backslash \{\mathbf{I}, \mathbf{E}, LHS(r_i)\}$
10:         create a fair data set for r_i by calling the function in Algorithm 5.2
11:         **if** $OR_{\mathbf{D}_f}(r_i) > m_{OR}$ or significantly greater than 1 **then**
12:             move r_i to **R**
13:             remove $LHS(r_i)$ from $\mathbf{X}_k$
14:         **end if**
15:     **end for**
16:     $k = k + 1$
17:     $\mathbf{X}_k = \text{Generate}(\mathbf{X}_{k-1})$
18:     $\mathbf{X}_k \leftarrow \text{Frequent}(\mathbf{X}_k)$
19: **end while**
20: output **R**

We briefly review the notation introduced in Chap. 4. The variables we refer to are binary variables, and a variable with value 1 indicates the presence of an event and 0 indicates the absence of the event. A combined variable is a composite consisting two or more variables and the value of the combined variable is the result of the logic AND operation on its component variables. An item (or an itemset) is the value 1 of a variable (or a combined variable). An item (or itemset) is frequent if its support is greater than the given minimum support. We use item (itemset) x and variable X interchangeably since we are only interested in the presence of an event. The function *Frequent()* in Algorithm 5.1 returns frequent items or itemsets.

When generating association rules, similar to CR-PA, the CR-CS algorithm also applies infrequent itemset pruning (Property 4.1) and uses a level by level approach to generating association rules and detecting causal rules from the association rules. As shown in Algorithm 5.1, in the k-th iteration (initially $k = 1$) of the **while** loop (Lines 5 to 19), each of the association rules of length k (generated in Line 6) is assessed to determine if it is a causal rule (Lines 7 to 15). Then the $(k+1)$-th level predictor variables are generated (Lines 17–18) by combining lower level itemsets and pruning infrequent itemsets. Apriori Gen [1] is used for generating combined itemsets at level $k > 1$.

Note that unlike CR-PA, CR-CS does not use Observation 4.2. The Observation is correct, but if an association test is unreliable, applying the Observation may lead to misses of true combined causal rules. In real world data sets, the number of combined causal rules may be small, so Observation 4.2 is not used by CR-CS to allow more candidate causal rules to be tested at the expense of efficiency.

In the following, we introduce the main steps for detecting a causal rule by validating an association rule, i.e. Lines 7–15 in Algorithm 5.1.

For an association rule r_i, in Line 9 of Algorithm 5.1, we select the set of control variables for constructing a fair data set for the rule. The control variable set includes all predictor variables excluding the LHS of r_i, irrelevant variables, exclusive variables, and variables with infrequent itemsets.

Irrelevant variables are identified in Line 3. They are the variables which are not associated with the target variable. For a variable X, its association with the target variable Z is determined by the odds ratio of $x \rightarrow z$. The significance of an association is tested using the approach described in Appendix A.

Exclusive variables of an exposure variable (a variable in the LHS of r_i) are identified in Line 8. They are the variables that have strong positive or negative associations with the exposure variable. If an exclusive variable is included in the control variable set, there is little chance for finding a fair data set. Please refer to Sect. 5.5.2 for an example.

Based on the control variable set, a fair data set is created for r_i (Line 10). Then the odds ratio of the rule on the fair data set is calculated using Formula (5.2). If the odds ratio is greater than the given minimum odds ratio, the rule is confirmed to be a causal rule and added to the output causal rule set (Line 12). The itemset of the LHS of a causal rule will not be considered in the next level for combining variables (Line 13).

The function in Algorithm 5.2 shows the procedure of creating a fair data set for a rule $x \rightarrow z$ based on a given data set $\mathbf{D}$ and the control variable set $\mathbf{C}$. In Line 2, $\mathbf{D}$ is split by values of X to two sub data sets, $\mathbf{D}_x$ and $\mathbf{D}_{\neg x}$ where records with $X = 1$ are in $\mathbf{D}_x$ and records with $X = 0$ are in $\mathbf{D}_{\neg x}$. In Lines 3–5, records in each of the two sub data sets are sorted by values of the control variables and a set of equivalence classes is formed in each sub data set, named as $\mathbf{G}_x$ and $\mathbf{G}_{\neg x}$, respectively. In Lines 6–7, pairs of matching equivalence classes in $\mathbf{G}_x$ and $\mathbf{G}_{\neg x}$ are identified such that in every two matching equivalent classes, the control variables have the same value. Since all pairs of records are matched in two matching equivalence classes, in Lines 8–11, pairs of records are randomly chosen from the two matching equivalence classes and moved to the fair data set $\mathbf{D}_f$. This selection repeats until one equivalence class is empty. The function terminates when all matching equivalence classes have been processed, and a fair data set $\mathbf{D}_f$ is returned.

Example 5.2. We use the data set in Table 5.2a to illustrate the CR-CS algorithm. The data set is generated based on the causal structure $B \rightarrow A \rightarrow Z$, with C being

---

Algorithm 5.2: Function: Creating a fair data set for rule $x \rightarrow z$

Input: Data set $\mathbf{D}$; rule $x \rightarrow z$; and the control variable set $\mathbf{C}$
Output: a fair data set for rule $x \rightarrow z$, $\mathbf{D}_f$
1: let $\mathbf{D}_f = \emptyset$
2: split $\mathbf{D}$ to $\mathbf{D}_x$ and $\mathbf{D}_{\neg x}$ where $X = 1$ in $\mathbf{D}_x$ and $X = 0$ in $\mathbf{D}_{\neg x}$
3: **for** $\mathbf{D}_x$ and $\mathbf{D}_{\neg x}$ each **do**
4:     sort data set w.r.t the values of $\mathbf{C}$ to form two sets of equivalence classes $\mathbf{G}_x$ and $\mathbf{G}_{\neg x}$
5: **end for**
6: **for** each equivalence class $\mathbf{E}_i$ in $\mathbf{G}_x$ **do**
7:     find a matching equivalence class $\mathbf{E}_j$ in $\mathbf{G}_{\neg x}$
8:     **while** ($\mathbf{E}_i \neq \emptyset$ and $\mathbf{E}_j \neq \emptyset$) **do**
9:         randomly pick up one record t_p in $\mathbf{E}_i$ and one record t_q in $\mathbf{E}_j$
10:         move matched records t_p and t_q from $\mathbf{E}_i$ and $\mathbf{E}_j$ to $\mathbf{D}_f$
11:     **end while**
12: **end for**
13: output $\mathbf{D}_f$

Table 5.2: **a** A summary of the original data set; **b** Matched pairs for testing $a \rightarrow z$; **c** Matched pairs for testing $b \rightarrow z$. Values of the control variables are highlighted. In both **b** and **c**, the left section includes the records of exposure group ($A = 1$) and the middle section includes the records of control group ($A = 0$). Matches are broken down to two equivalence classes. The right section (#matched pairs column) shows the number of matched pairs

A B C Z	count
0 0 0 0	61
0 0 0 1	18
0 0 1 0	168
0 0 1 1	28
0 1 0 0	39
0 1 1 0	100
0 1 1 1	24
1 0 0 0	15
1 0 0 1	47
1 0 1 0	21
1 0 1 1	111
1 1 0 1	81
1 1 1 0	57
1 1 1 1	191

(a)

matched pairs		
A B Z	A B Z	#matched pairs
1 1 1	0 1 1	18
1 1 1	0 1 0	115
1 1 0	0 1 1	6
1 1 0	0 1 0	24
1 0 1	0 0 1	26
1 0 1	0 0 0	132
1 0 0	0 0 1	8
1 0 0	0 0 0	28

(b)

matched pairs		
B A Z	B A Z	#matched pairs
1 1 1	0 1 1	130
1 1 1	0 1 0	30
1 1 0	0 1 1	28
1 1 0	0 1 0	6
1 0 1	0 0 1	4
1 0 1	0 0 0	20
1 0 0	0 0 1	24
1 0 0	0 0 0	115

(c)

independent of A, B and Z. Let the minimum support be 48, approximately 5 % of the total number of records, and the minimum odds ratio be 1.5.

The data set is the same as the data set in Example 4.1. Readers can refer to Table 4.3 for the itemised data view.

Since supp(az) = 430 > 48, supp(bz) = 296 > 48 and supp(cz) = 354 > 48, no variable is removed and the current variable set $\mathbf{X} = \{A,B,C\}$.

CR-CS tests three candidate association rules $a \to z$, $b \to z$ and $c \to z$ at the first level. Their contingency tables are:

	z	$\bar{z}$	Total
a	430	93	523
$\bar{a}$	70	368	438
Total	500	461	961

	z	$\bar{z}$	Total
b	296	196	492
$\bar{b}$	204	265	469
Total	500	461	961

	z	$\bar{z}$	Total
c	354	346	700
$\bar{c}$	146	115	261
Total	500	461	961

We have $OR_{\mathbf{D}}(a \to z) = 24.3$, $OR_{\mathbf{D}}(b \to z) = 1.96$ and $OR_{\mathbf{D}}(c \to z) = 0.81$. Rules $a \to z$ and $b \to z$ are association rules, and rule $c \to z$ is not. C is an irrelevant variable since $c \to z$ is not an association rule. Irrelevant variable set $\mathbf{I} = \{C\}$, and C will not be in the control variable set of $a \to z$ or $b \to z$.

We firstly generate a fair data set for rule $a \to z$. B is not an exclusive variable of A. So the control variable for rule $a \to z$ is B. A fair data set of $a \to z$ is shown in Table 5.2b. The matched pairs and their counts are listed for easy calculation of odds ratio. Note that matching is a random process and hence the table just lists one possible fair data set. In experiments, we should repeat the test a number of times and choose the rule by the majority votes. In this example, only one trial is shown.

Based on the fair data set for $a \to z$, we have $n_{12}^f = 115 + 132 = 247$ and $n_{21}^f = 6 + 8 = 14$. $OR_{\mathbf{D}_f}(a \to c) = 17.6$, which is greater than 1.5, the user specified minimum odds ratio. Therefore, $a \to z$ is a causal rule.

We then generate a fair data set for rule $b \to z$. A is not an exclusive variable of B. So the control variable for rule $b \to z$ is A. A fair data set of $b \to z$ is listed in Table 5.2c.

Based on the fair data set for $b \to z$, we have $n_{12}^f = 30 + 20 = 50$ and $n_{21}^f = 28 + 24 = 52$. $OR_{\mathbf{D}_f}(b \to z) = 0.96$. Therefore, rule $b \to z$ is not a causal rule.

Now the variable set $\mathbf{X} = \{B,C\}$ since A is removed ($a \to z$ is a causal rule). The only candidate generated at Level 2 is $bc \to z$. The contingency table of rule $bc \to z$ is:

	z	$\bar{z}$	Total
bc (both $B = 1$ and $C = 1$)	215	157	372
$\overline{bc}$ (either $B = 0$ or $C = 0$)	285	304	589
Total	500	461	961

From the table, $OR_{\mathbf{D}}(bc \to z) = 1.46 < 1.5$, hence $bc \to z$ is not an association rule. The algorithm is terminated, and rule $a \to z$ is returned.

In this example, we know that the association between B and Z is intermediated by variable A. So B is not a cause of Z. This result is consistent with the underlying causal model.

5.5 Discussions

5.5.1 Complexity of CR-CS

The time complexity of CR-CS mainly attributes to the two tasks: association rule mining and fair data set construction.

The time complexity of association rule mining has been analysed in the previous chapter, and we do not repeat the analysis here. Note that there are a lot of efficient association rule mining algorithms [5].

The construction of a fair data set for an association rule takes the most time for a simulated cohort study. For each rule, a data set is split by the values of the exposure variable into two sub data sets. This takes $O(n)$ time where n is the number of records in a data set. Then in each sub data set, records are sorted by values of the control variables. This takes $O(n \log n)$ time. Suppose that p equivalence classes are formed in each sub data set, finding a pair of matching equivalence classes in the two sub data sets is $O(p)$ since equivalence classes in both sub data sets have been sorted and the matches can be found by an insertion sort. In each pair of equivalence classes, the time taken for finding the matched pairs of records is $O(n/p)$ because matched pairs are randomly picked up. Since $p \ll n$, we may ignore $O(p)$. In comparison with $O(n \log n)$, $O(n/p)$ can be ignored. The overall complexity for constructing a fair data set is $O(n \log n)$.

The complexity for testing causal rules is $|\mathbf{R}_a| O(n \log n)$ where $|\mathbf{R}_a|$ is the number of association rules. When we do not consider high level combined variables with many items in the LHS of a rule, the number of association rules will not be exponential to the number of variables, hence the algorithm is efficient.

If we are only interested in causal rules with single item in their LHS, CR-CS is very efficient. In this case, the number of association rules is at most m (the number of predictor variables), and based on the above analysis of the time taken for a simulated cohort study, the time complexity of CR-CS is $O(mn \log n)$.

5.5.2 False Discoveries of CR-CS

For CR-CS, we also discuss the false discoveries from two aspects: algorithm design and the input data.

Using a set of control variables (instead of all other variables) in constructing a fair data set for a rule is necessary for CR-CS. For example, for 30 control variables, we will need $2 * 2^{30} = 2147483648$ data records to have two records for each level (value assignment) of the control variables. Even a big data set is only able to support tens of control variables.

The choice of control variables is a major cause of false discoveries in CR-CS. Generally speaking, the more control variables, the fewer false positives but the more false negatives. When we control all other variables we have observed, if a causal relationship is identified, it is unlikely false. However, this is impossible for most data sets. When we control many variables, true causal rules may not be found because of data insufficiency. In contrast, the fewer the control variables, the more false positives. For example, if we do not control a variable that is a common cause of the exposure variable and the target variable, a false discovery will be introduced. So the determination of a right set of control variables is a major challenge for CR-CS (and CR-PA).

Using relevant variables (those associated with the target variable) as control variables is conceptually right. However, when the number of relevant variables is big, we still have to choose only some of them. The set of control variables determines the size of a fair data set. If the control variable set is large, it is likely to find an empty fair data set, and this makes a cohort study impossible. However, using too few control variables will lead to false discoveries since an outcome may be caused by an uncontrolled variable instead of the exposure variable. The choice is a balance between having a non-empty fair data set and achieving quality causal discovery.

Furthermore, the combination of multiple irrelevant variables may be relevant. However, CR-CS does not use combined variables as control variables, as the number of combined variables is big (hence combined relevant variables may be many) and the support of a combined variable (i.e. the number of records containing value 1 of all component variables) is normally small. If they are included in the control variable set, it is very likely to have empty exposure or control groups. However, not controlling combined variables which are causes of the target will lead to false positives.

The exclusion of exclusive variables from the control variable set is a practical solution since we could not find a causal relationship when a control variable is highly positively or negatively associated with the exposure variable. We use the examples in Table 5.3 to show the limitation by involving an exclusive variable in the control variable set. In Table 5.3, $a \to z$ is a causal rule. If the control variable set includes variable D that is negatively associated with variable A, the fair data set will be empty since there is no record matching between groups $A = 1$ and $A = 0$. If the control variable set includes variable E that is positively associated with variable A. The fair data set will be empty since no records match between groups $A = 1$ and $A = 0$. Causal rule $a \to z$ is unidentifiable if a variable like D or E is used in the control variable set. However, if D or E is excluded, in some cases $a \to z$ may be identified as a false positive, for example, when E is a common cause of A and Z.

False discoveries will be produced when the input data does not satisfy the assumptions made for the algorithm and the requirements for the statistics test. Firstly, the data should be free of selection bias to ensure the correctness of CR-CS. CR-CS also assumes causal sufficiency (that all causal variables have been measured) and valid statistical tests. Causal sufficiency is normally not satisfied since we cannot guarantee that all cause variables are measured. So any discovery is based on the currently available information. Secondly, the odds ratio tests require that the counts in a contingency table are not too small. Normally, the count in each cell is five or greater. When the counts in contingency tables are small, false discoveries will occur. Thirdly, because of the randomness of choosing matched pairs, it is nec-

Table 5.3: Examples showing the effects of exclusive variables. **a** D and A are negatively associated; **b** E and A are positively associated. Values of control variables are highlighted. There are no matches for the values of control variables between $A = 1$ and $A = 0$ groups in both cases

A	D	B	C	Z
0	1	0	0	0
0	1	0	0	0
0	1	0	1	0
0	1	0	1	0
0	1	1	0	0
0	1	1	1	0
1	0	0	0	1
1	0	0	0	1
1	0	0	1	1
1	0	0	1	1
1	0	1	1	1
1	0	1	1	1

(a)

A	E	B	C	Z
0	0	0	0	0
0	0	0	0	0
0	0	0	1	0
0	0	0	1	0
0	0	1	0	0
0	0	1	1	0
1	1	0	0	1
1	1	0	0	1
1	1	0	1	1
1	1	0	1	1
1	1	1	1	1
1	1	1	1	1

(b)

essary to run CR-CS a number of times to obtain consistent results. Finally, CR-CS needs a large data set to obtain a reliable fair data set.

5.6 An Implementation of CR-CS

We assume that readers have installed the CRE software tool in their computers by following the instructions in Chapter 4. In the following, we demonstrate how to use CRE to discover causal rules by CR-CS.

5.6.1 Example 1: Running CR-CS in CRE

With this example, we use the data set discussed in Example 5.2, which can be downloaded from *http://nugget.unisa.edu.au/Causalbook/*. The input files are in C4.5 format, including *Example41.data* and *Example41.names*. The following steps demonstrate the procedure of running CR-CS in CRE.

1. Open the input file. Select on the menu bar, File → Open File. Select the input *.names* file from local computer, in this case, select *Example41.names*.
2. Select the algorithm CR-CS from the left panel.
3. Set parameters for CR-CS. On the menu bar, select Configuration → Parameters. A pop-up window will appear for setting CR-CS parameters as shown in Fig. 5.1. There are three parameters as follows:

- *Max level of combined rules*: l_{max} in Algorithm 5.1, which is the maximum level of combined causes that users desire, e.g. select 2 for mining single and combined (level 2) causal rules.
- *MinSupport*: the minimum support, i.e. m_{supp} in Algorithm 5.1.
- *Odds ratio*: the odds ratio threshold for mining association rules and causal rules. The default value is "lower bound", which uses the lower bound of the confidence interval of the odds ratio as the criterion (see Appendix A for more details)

Click "Confirm" after all of the parameters are set.

Fig. 5.1: The interface for parameter setting in CR-CS

4. Select attributes from the data set. Click the "Choose Attributes" button in the middle panel of the CRE user interface to select the attributes (predictor variables) of interest as shown in Fig. 5.2. Click the "All" button to select all attributes in the data set, or alternatively tick the boxes to choose a subset of attributes. Note that the target variable, which is the last column in the data set, is not included in the list of attributes. Click "Confirm" after selecting the desired attributes.
5. Run CR-CS. Click the "Start" button to run the CR-CS algorithm.
6. Obtain the results. The outputs are shown in the CR-CS output panel as in Fig. 5.3. Moreover, the results are also stored in the file *current_optforUI.csv*, which is located in the current folder of *CRE*. Each row of the report is a rule. The "Causal/Noncausal" column tells us if the rule is a causal rule or it is just an association rule. The "level" column shows the length of the LHS of a rule, with level 1 indicating single variable in the LHS of a rule and level 2 indicating 2 variables in LHS of a rule. The last four columns show the statistics of the rule which are the values of the four cells in the contingency table. We can see from the result that there is only one causal rule at level one, $A = 1 \rightarrow Z = 1$, which is consistent with the causal structure $B \rightarrow A \rightarrow Z$ in Example 5.2.

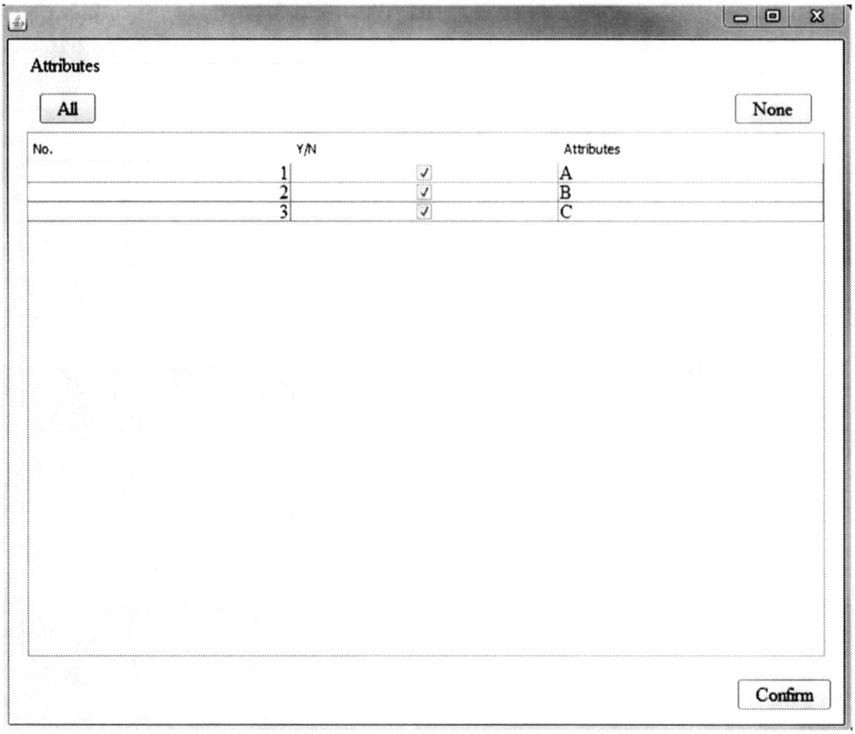

Fig. 5.2: The interface for choosing attributes in CR-CS

5.6.2 Example 2: Using the Data Sets of Figs. 2.1 and 2.2

The data set for the Bayesian network shown in Fig. 2.1 in Chap. 2 can be down-loaded from: *http://nugget.unisa.edu.au/Causalbook/.* This data set is the same the one used in Chaps. 2 and 3, but it has been converted into the C.4.5 format to have two files *Example21.data* and *Example21.names.*

Using the same procedure as in the above example, we set the maximum level of combined variables to 2, and keep the default setting for other parameters as in Fig. 5.1. The result is shown in Fig. 5.4.

We can see from the result that B, C, and F have causal relationships with the target Z, which is consistent with the results from PC-simple, HITON-PC, and CR-PA. For example, at the first level, we have $C = 1 \rightarrow Z = 1$, $B = 0 \rightarrow Z = 1$, and $F = 0 \rightarrow Z = 1$, which are consistent with the results of CR-PA. However, CR-CS produces rules with both target values, $Z = 1$ and $Z = 0$. Therefore, apart from the mentioned rules with target $Z = 1$, we also have $C = 0 \rightarrow Z = 0$, $B = 1 \rightarrow Z = 0$, and $F = 1 \rightarrow Z = 0$.

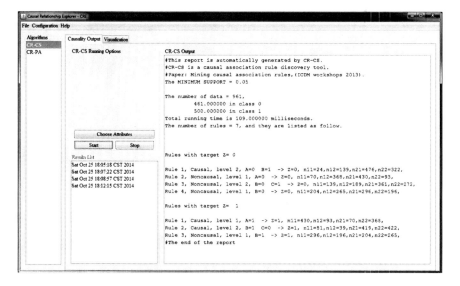

Fig. 5.3: The output of CR-CS algorithm for the data set in Example 5.2

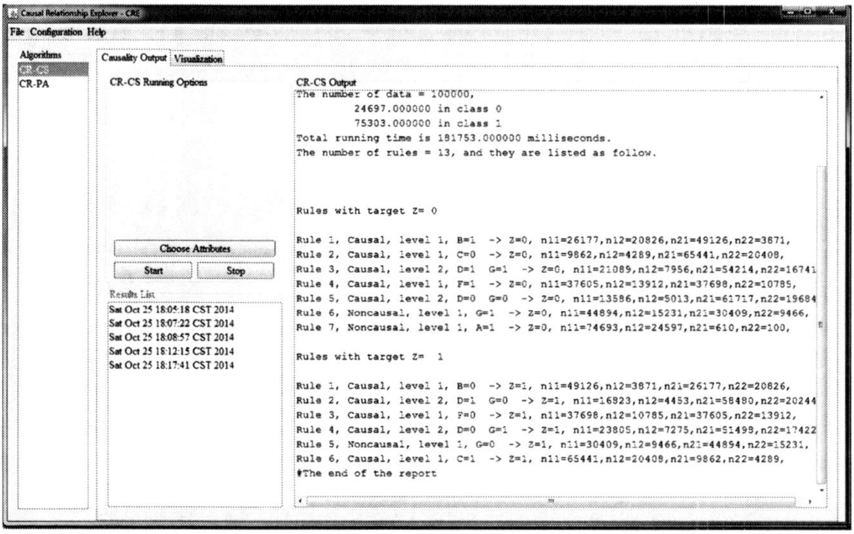

Fig. 5.4: The output of CR-CS algorithm for the data set in Fig. 2.1

Similarly, we apply CR-CS to the data set of Fig. 2.2 which can be downloaded from: *http://nugget.unisa.edu.au/Causalbook/*. We use the same settings as above, and the result is shown in Fig. 5.5.

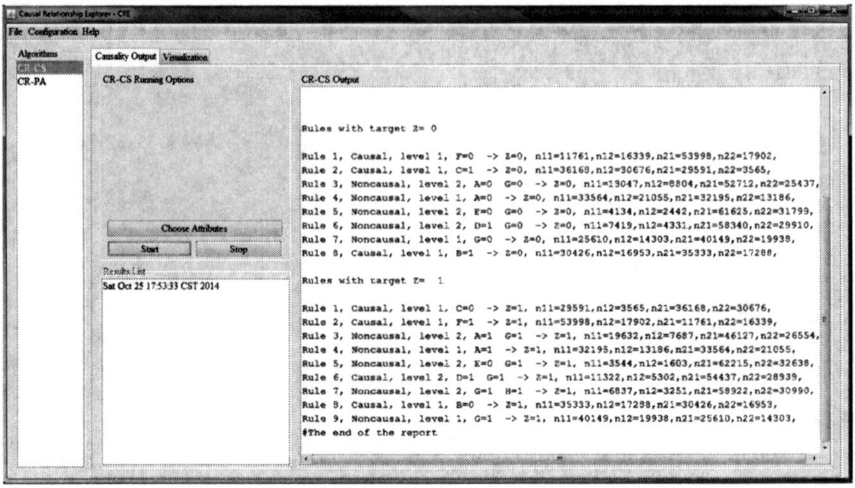

Fig. 5.5: The output of CR-CS algorithm for the data set in Fig. 2.2

With this data set, CR-CS discovers the parents and children set of Z as $\{B, C, F\}$ which is different from the results of PC-simple, HITON-PC and CR-PA. The set of single causal rules is the same for both data sets of Figs. 2.1 and 2.2.

References

1. R. Agrawal, H. Mannila, R. Srikant, H. Toivonen, and A. I. Verkamo. Fast discovery of association rules. In *Advances in Knowledge Discovery and Data Mining*, volume 12, pages 307–328. AAAI/MIT Press Menlo Park, CA, 1996.
2. J. Concato, Shah N, and R. I. Horwitz. Randomized, controlled, trials, observational studies, and the hierarchy of research design. *The New England Journal of Medicine*, 342(25):1887–1892, June 2000.
3. A. M. Euser, C. Zoccali, K. Jager, and F. W. Dekker. Cohort studies: prospective versus retrospective. *Nephron Clinical Practice*, 113:214–217, 2009.
4. J. L. Fleiss, B. Levin, and M. C. Paik. *Statistical Methods for Rates and Proportions*. Wiley, 3rd edition, 2003.
5. J. Han and M. Kamber. *Data Mining: Concepts and Techniques*. Morgan Kaufmann Publishers Inc., San Francisco, CA, USA, 2nd edition, 2005.
6. J. Li, T. D. Le, L. Liu, J. Liu, Z. Jin, and B. Sun. Mining causal association rules. In *ICDM Workshops*, pages 114–123, 2013.
7. P. R. Rosenbaum. *Design of Observational Studies*. Springer Series in Statistics. Springer, 2010.
8. J. W. Song and K. C. Chung. Observational studies: Cohort and case-control studies. *Plastic & Reconstructive Surgery*, 126(6):2234–2242, December 2010.
9. E. A. Stuart. Matching methods for causal inference: a review and a look forward. *Statistical Science*, 25(1):1–21, 2010.

Chapter 6
Experimental Comparison and Discussions

Abstract In Chaps. 2–5 we have presented four methods for discovering the causal relationships around a given target variable, focusing on the basic ideas and procedures of the methods. In this chapter, we apply the methods to a series of data sets to evaluate and compare their efficiency. Discussions on the experimental results are provided in this chapter too.

6.1 Data Sets and Settings

Sixteen synthetic data sets are employed in the experiments. To generate the data sets, we firstly create Bayesian network structures where some predictor variables are parents or children of the target variable, and some are not. We then simulate the binary data sets of these Bayesian networks using the logistic regression model, which is the same as the one introduced in the *pcalg* package [1]. For all data sets, we set the number of parents and children of the target in the range of 10–20 % of the total number of variables.

A summary of the data sets is given in Table 6.1. The number of variables in the table refers to the number of variables, including all predictor variables and the target variable in a data set. All predictor variables and the target variable are binary variables. The distributions column refers to the percentages of the two different values of the target variable in the data sets. The last column, $|\mathbf{PC}|$, shows the number of parents and children of the target.

The first five synthetic data sets are used to compare the efficiency of the algorithms, and the rest of the synthetic data sets are used to investigate the scalability of the algorithms in terms of the number of variables and samples.

In the experiments, we set the default minimum support for CR-PA and CR-CS to 0.05. The minimum odds ratio for causal rules in CR-CS is set to 1.5 and the confidence level for Chi-square tests in CR-PA is set to 95 %. Meanwhile, the significance level for conditional independence tests in PC-simple and HITON-PC is set to 0.05.

© The Author(s) 2015   67
J. Li et al., *Practical Approaches to Causal Relationship Exploration*,
SpringerBriefs in Electrical and Computer Engineering, DOI 10.1007/978-3-319-14433-7_6

Table 6.1: A summary of data sets used in experiments

Name	#Records	#Variables	Distributions	|PC|
Syn50-5K	5000	50	19.8 % & 80.2 %	10
Syn60-5K	5000	60	19.8 % & 80.2 %	13
Syn80-5K	5000	80	19.8 % & 80.2 %	18
Syn100-5K	5000	100	19.8 % & 80.2 %	20
Syn120-5K	5000	120	19.8 % & 80.2 %	26
Syn200-10K	10,000	200	19.8 % & 80.2 %	20
Syn400-10K	10,000	400	19.8 % & 80.2 %	40
Syn600-10K	10,000	600	19.8 % & 80.2 %	60
Syn800-10K	10,000	800	19.8 % & 80.2 %	80
Syn1000-10K	10,000	1000	19.8 % & 80.2 %	100
Syn100-10K	10,000	100	19.8 % & 80.2 %	20
Syn100-20K	20,000	100	19.8 % & 80.2 %	20
Syn100-40K	40,000	100	19.8 % & 80.2 %	20
Syn100-60K	60,000	100	19.8 % & 80.2 %	20
Syn100-80K	80,000	100	19.8 % & 80.2 %	20
Syn100-100K	100,000	100	19.8 % & 80.2 %	20

All experiments were conducted on a computer with 4 core Intel i7-3370 CPU @ 3.4 GHz and 16 GB RAM running 64-bit Windows operating system.

6.2 Efficiency of Methods

The running time (in seconds) of all methods with the first five synthetic data sets is listed in Table 6.2. Overall, for finding single causes, CR-PA is the most efficient method and PC-simple is the worst. We can see from Table 6.2 that PC-simple is only suitable for data sets with a small number of variables since its time complexity is exponential to the number of predictor variables in the worst case. It took 2 days for PC-simple to complete with the data set with 120 nodes (26 parents and children). In theory, the quality of discoveries by PC-simple is high. CR-PA and HITON-PC are fast. CR-CS is slower than CR-PA and HITON-PC, but it is reasonably fast.

In terms of the running time for generating combined causal rules, CR-PA is overall more efficient than CR-CS with small data sets, but the running time of CR-PA increases quickly when the number of variables in the data set is increased. This is because the numbers of candidate association rules approximate to the same for both CR-CS and CR-PA with the increase of irrelevant variables.

To evaluate the scalability of the methods in terms of the number of records, we apply the four methods to the last six synthetic data sets (Syn100-10K, Syn100-20K ..., Syn100-100K) in Table 6.1. The data sets have the same number of variables (100), but the number of records varies from ten thousands to one hundred thousands. For CR-PA and CR-CS, we ran them in two different versions, respectively.

Table 6.2: Running time (seconds) of the methods

	Syn50-5K	Syn60-5K	Syn80-5K	Syn100-5K	Syn120-5K
PC-simple	1	8.7	430.4	1987.8	205764.9
HITON-PC	0.6	0.9	1.1	1.7	2.5
CR-PA (single)	0.4	0.5	1.0	1.1	1.5
CR-CS (single)	4.1	5.1	8.1	9.4	13.6
CR-PA (combined)	5.1	8.2	25.2	44.0	66.8
CR-CS (combined)	13.9	15.6	28.3	43.1	55.9

With CR-PA1 and CR-CS1, we constrained the length of rules as 1, to make them comparable with PC-simple and HITON-PC. With CR-PA2 and CR-CS2, the length of the rules was restricted to 2 to allow the discovery of combined cause factors.

The execution time (in seconds) of the methods with respect to the number of records in the data sets is shown in Fig. 6.1. From the figure, we can see that PC-simple is the slowest method. Other methods are quite fast when the number of records varying from 10K to 100K. Both CR-PA1 and CR-CS1 are very efficient, but CR-PA2 is not scalable, and so is CR-CS2 when the number of samples is increased. HITON-PC performs similarly as CR-PA1 and CR-CS1 for the first four synthetic data sets (10K, 20K, 40K, and 60K), but its running time increases more quickly than CR-PA and CR-CS for the last two data sets (80K and 100K). Generally speaking, they are all scalable with the size of data sets (number of samples).

We also apply the methods to five synthetic data sets with the same number of records but different number of variables (Syn200-10K, Syn400-10K, Syn600-10K, Syn800-10K, Syn1000-10K) to assess the scalability of the methods with respect to the number of variables. Figure 6.2 shows the running time of PC-simple, HITON-PC, CR-PA1, CR-PA2, CR-CS1, and CR-CS2. As expected, PC-simple is inefficient and the running time quickly becomes impractical when the number of variables is increased. CR-CS2 and CR-PA2 are also inefficient when the number of variables is large. When the number of variables is increased, a lot of combined causal rules need to be tested and therefore both methods require long running time at the second level. It took more than four hours for CR-PA2 and CR-CS2 to complete with the data sets with 600 variables (Syn600-10K). Meanwhile, CR-CS1, CR-PA1 and HITON-PC are very efficient and scalable in terms of the number of variables.

6.3 Discussions

Table 6.3 gives the suitability of different methods for different data types. With low dimensional data sets (the number of variables is small), theoretically all the methods are suitable except that CR-CS requires sufficient samples for generating proper fair data sets. We suggest using PC-simple with low dimensional data sets, especially when the number of samples is not too large, since it is expected to pro-

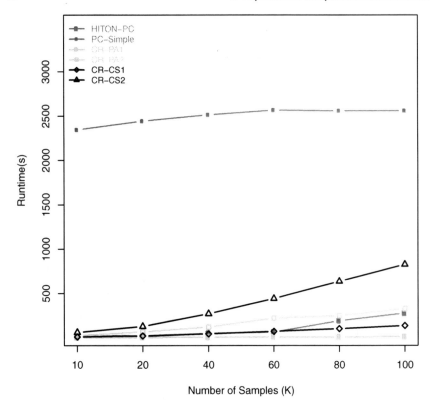

Fig. 6.1: Scalability of the methods with data size

vide quality discoveries. When the number of samples is large, PC-simple may need very long running time due to conditional independence tests.

Table 6.3: Recommendation of methods to use

	Low dimensional	High dimensional
Large data	HITON-PC, CR-PA, CR-CS	HITON-PC, CR-PA, CR-CS
Small data	PC-simple	HITON-PC

With high dimensional data sets, HITON-PC should work for both large and small data sets. For a small data set, max_k can be set to a very small value since small data sets only support low order conditional interdependence tests (though the quality of discoveries may be traded off). In contrast, CR-PA and CR-CS only work on the data sets with a lot of samples since otherwise no reliable statistics will be produced. As discussed in Chap. 3, HITON-PC is more efficient when we set a low max_k, but the accuracy will be traded off for the high efficiency. Similarly, we can also set the maximum number of control variables for CR-PA and CR-CS to make

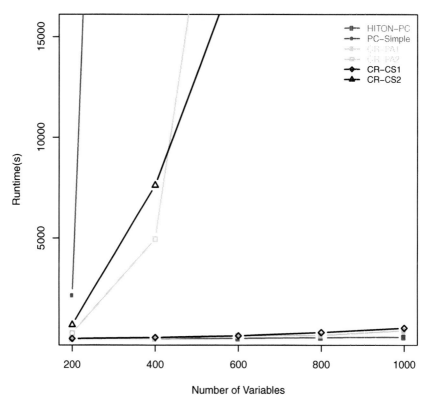

Fig. 6.2: Scalability of the methods with the number of variables

them scalable with large and high dimensional data sets, but the output causal rules in this case may involve false discoveries.

For high dimensional data sets with a small number of records, the methods discussed in this book may not produce reliable results when the underlying causal structure is dense, i.e. the parents and children set of the target variable is large. The statistical tests in PC-simple and HITON-PC require sufficiently large number of records to be reliable. CR-PA also requires a large number of samples for testing partial association when the number of control variables is large. CR-CS may not produce any causal rules in this setting, as the size of the fair data sets would be very small.

On another note, the choice of association / dependence tests is important for the success of an algorithm for inferring causal relationships especially when the number of samples is not sufficiently large. A few different criteria being introduced with the different methods in this book do not indicate that they are suitable for these methods, but that they are associated with the implementations of the methods. The accuracy of association tests depends on the satisfaction of assumptions. For example, Z-Fisher dependence test assumes linearity and normality. When the assumptions are not satisfied, false positives and false negatives will be produced. Some assumptions may be relaxed and some may not [2].

In most cases, when a method does not produce accurate relationships, this is not because the method is not good, but the data set does not satisfy the assumptions of the association/dependence tests. Theoretically, we may firstly test the satisfaction of various assumptions. However, it is difficult in practice since such a test again relies on assumptions.

In practice, it is helpful to have a data set with (some or all) known causal relationships. So users can use different association / dependence tests to produce causal relationships and compare them with the known truths. In this way, users can choose a right association test criterion.

References

1. D. Colombo, A. Hauser, M. Kalisch, and M. Maechler. Package '*pcalg*', 2014.
2. V. Mark and D.J. Marek. Insensitivity of Constraint-Based Causal Discovery Algorithms to Violations of the Assumption of Multivariate Normality. In *The Florida AI Research Society Conference*, pages 690–695, 2008.

Appendix A
Appendix

In statistics, the term association is used to represent a relationship between two random variables which are statistically dependent and the relationship is not necessary to be causal. To measure the strength of the association between two variables, statisticians have developed several methods, in which the most popular ones are correlation test methods where the relationship between two variables is assumed to be linear. In this book, we do not distinguish association and correlation and use the two terms exchangeably. In the following we introduce the definitions of independence, association and correlation, as well as the methods to test those relationships.

A.1 Tests of Independence and Conditional Independence

In this section, we introduce the statistical tests for detecting an association between two binary variables as well as the conditional independence between two binary variables conditioning on a set of variables, including Fisher's z-transform, Chi-square, and odds ratio. These tests are used by the causal discovery methods presented in the book.

Before discussing the tests for associations, we firstly present the definition of the independence and conditional independence relationship between two variables.

Definition A.1 (Independent and conditionally independent variables). Two variables, X and Z are independent if whenever $P(Z = z) > 0$ we have:

$$P(X = x, Z = z) = P(X = x)P(Z = z)$$

where x and z are possible values of X and Z respectively, e.g. $x = 1$ or $x = 0$ for binary variable X. If X and Z are not independent, they are also called associated variables. Two variables X and Z are conditionally independent given a set of variables $\mathbf{S}$, denoted as $\mathrm{Ind}(X, Z|\mathbf{S})$, if $P(X = x, Z = z|\mathbf{S} = s) = P(X = x|\mathbf{S} = s)P(Z = z|\mathbf{S} = s)$ for all the values of X, Z, and $\mathbf{S}$, such that $P(\mathbf{S} = s) > 0$.

© The Author(s) 2015
J. Li et al., *Practical Approaches to Causal Relationship Exploration*,
SpringerBriefs in Electrical and Computer Engineering, DOI 10.1007/978-3-319-14433-7

A.1.1 Fisher's z-Transform for Correlation and Partial Correlation

The PC-simple method presented in Chap. 2 uses Pearson's correlation coefficient (Pearson's r) or partial correlation to measure an association and utilises Fisher's z-transformation or Fisher's r-to-z transformation to assess the strength of a correlation.

Given a data set containing n samples of variables X and Y, the Pearson's r (sample correlation coefficient) is calculated as:

$$r = \frac{\sum_{i=1}^{n}(X_i - \bar{X})(Y_i - \bar{Y})}{\sqrt{\sum_{i=1}^{n}(X_i - \bar{X})^2}\sqrt{\sum_{i=1}^{n}(Y_i - \bar{Y})^2}}$$

where $\bar{X}$ and $\bar{Y}$ are the means of X and Y, respectively.

Fisher's r-to-z transformation converts the Pearson's coefficients to the normally distributed variable, $z \sim N(\mu_z, \sigma_z)$ using the formula:

$$z = 0.5\ln\left(\frac{1+r}{1-r}\right)$$

and

$$\mu_z = 0.5\ln\left(\frac{1+\rho}{1-\rho}\right), \quad \sigma_z = \frac{1}{\sqrt{n-3}}$$

where ρ is the population Pearson correlation coefficient.

With a given significance level α and a calculated r value, we conclude X and Y are dependent if:

$$z\sqrt{n-3} > \phi^{-1}(1-\alpha/2)$$

where ϕ is the standard Normal cumulative distribution function.

For example, assuming that from a data set of X and Y with 100 samples, we have calculated the Pearson's coefficient r as 0.4. For a 95 % confidence level (significance level $\alpha = 0.05$), we have:

$$z\sqrt{n-3} = 0.5\ln\left(\frac{1+0.4}{1-0.4}\right)\sqrt{100-3} = 4.172458$$

and

$$\phi^{-1}(1-\alpha/2) = \phi^{-1}(1-0.05/2) = \phi^{-1}(0.975) = 1.959964$$

where the value of $\phi^{-1}(0.975)$ can be obtained by using a table of the standard Normal cumulative distribution function or using the function $qnorm(0.975)$ in R. As $z\sqrt{n-3} > \phi^{-1}(1-\alpha/2)$, we conclude that X and Y are dependent or associated.

Similarly, given the value of the partial correlation, Fisher's z-transform can be used to assess the conditional independence between two variable conditioning on a set of variables. Given a data set containing n samples of variables the partial correlation between X and Y given a set of variables $\mathbf{S}$ is calculated recursively as:

$$r_{X,Y|\mathbf{S}} = \frac{r_{X,Y|\mathbf{S}\setminus W} - r_{X,W|\mathbf{S}\setminus W} r_{Y,W|\mathbf{S}\setminus W}}{\sqrt{(1 - r^2_{X,W|\mathbf{S}\setminus W})(1 - r^2_{Y,W|\mathbf{S}\setminus W})}}$$

where W is a variable (any) in $\mathbf{S}$

The implementation of this recursive partial correlation formula is inefficient. An efficient implementation that uses James–Stein-type shrinkage estimator for the covariance matrix can be found in the R package called *corpcor* [2].

With a given significance level α, X and Y are dependent given $\mathbf{S}$ if:

$$z\sqrt{n - |\mathbf{S}| - 3} > \phi^{-1}(1 - \alpha/2)$$

where $|\mathbf{S}|$ is the length of $\mathbf{S}$, i.e. the number of variables in $\mathbf{S}$, and

$$z = 0.5\ln\left(\frac{1 + r_{X,Y|\mathbf{S}}}{1 - r_{X,Y|\mathbf{S}}}\right)$$

A.1.2 Chi-Square Test of Independence and Conditional Independence

Both HITON-PC (Chap. 3) and CR-PA (Chap. 4) use Chi-square test of independence for association test. HITON-PC also uses Chi-square test of conditional independence.

With two binary variables X and Z, given the contingency table summarising the data set and the significance level α, Chi-square test of independence will reject or fail to reject the null hypothesis that X and Z are independent.

A contingency table is a cross table showing the frequencies of the joint data distribution of two variables. The following table gives the contingency table of two binary variables X and Z. In this contingency table n is the size of the data

Table A.1: The 2×2 contingency table

	$Z = 1$	$Z = 0$	Total
$X = 1$	n_{11}	n_{12}	n_{1*}
$X = 0$	n_{21}	n_{22}	n_{2*}
Total	n_{*1}	n_{*2}	n

set, i.e. number of samples. n_{11}, n_{12}, n_{21}, and n_{22} are counts of $(X = 1, Z = 1)$, $(X = 1, Z = 0)$, $(X = 0, Z = 1)$, and $(X = 0, Z = 0)$ in the data set, and they are called observed frequencies. n_{1*}, n_{2*}, n_{*1}, and n_{*2} are counts of $X = 1, X = 0, Z = 1$ and $Z = 0$, and they are called marginal frequencies.

Based on the contingency table, we can calculate the Chi-square test statistic as follows.

$$\chi^2 = \sum_{i=1}^{i=2} \sum_{j=1}^{j=2} \frac{(n_{ij} - \hat{n}_{ij})^2}{\hat{n}_{ij}}$$

where $\hat{n}_{11}$, $\hat{n}_{12}$, $\hat{n}_{21}$, and $\hat{n}_{22}$ are the expected frequencies and they are calculated as:

$$\hat{n}_{ij} = \frac{n_{i*}n_{*j}}{n}$$

For the specified significance level α, we can consult the Chi-square distribution (degree of freedom $df = 1$) to find out the corresponding critical value χ_α^2. If the calculated Chi-square test statistic is greater than χ_α^2, we reject the null hypothesis and consider that X and Z are associated at the given significance level. Alternatively, we can consult the Chi-square distribution to find out the p-value of the obtained test statistic. If the p-value is less than the significance level, we reject the null hypothesis and accept the association.

For example, with the following contingency table for variables Female and Low Salary where F=1 is for female and F = 0 for male; LS = 1 for low salary and LS = 0 otherwise, the Chi-square test statistic obtained from this table is 2.73 (and the p-value is 0.099). When $\alpha = 0.05$, the critical value is 3.84. Because the test statistic is less than the critical value (or the p-value is greater than α), we fail to reject the null hypothesis and consider that Female and Low Salary are not associated.

Table A.2: An example 2×2 contingency table

	LS = 1	LS = 0	Total
F = 1	65	60	125
F = 0	185	120	305
Total	250	180	430

In the case of conditional independence test, e.g. testing the association between X and Z conditioning on a set of variables $\mathbf{S}$, we create the contingency tables for X and Z given each value of S. For example, if $\mathbf{S}$ contains a single variable W, then we have two contingency tables corresponding to $W = 1$ and $W = 0$ as follows.

Table A.3: The contingency tables with $W = 1$ and $W = 0$

$W = 1$	$Z = 1$	$Z = 0$	Total
$X = 1$	n_{11}	n_{12}	n_{1*}
$X = 0$	n_{21}	n_{22}	n_{2*}
Total	n_{*1}	n_{*2}	n

$W = 0$	$Z = 1$	$Z = 0$	Total
$X = 1$	n_{11}	n_{12}	n_{1*}
$X = 0$	n_{21}	n_{22}	n_{2*}
Total	n_{*1}	n_{*2}	n

The Chi-square value of a conditional independence test is the sum of all the Chi-square values in all contingency tables. For example, if the Chi-square value is 1.7 from the contingency table with $W = 1$, and it has the value of 0.9 from the table with $W = 0$, then the final Chi-square value is $1.7 + 0.9 = 2.6$. Similar to the unconditional dependence test, we can conclude the dependence or independence between X and Z by comparing the Chi-square value with the χ^2_α. Note that the degree of freedom for conditional independence test is $2^{|S|}$, where $|S|$ is the length of the conditioning set, i.e. the number of variables in S.

In Chap. 4, we test the dependence not only between a single variable and the target variable, but also between the combined variables and the target. The combined variable is considered as a new single variable whose value is 1 if all of component variables are 1, and it receives the value of 0 otherwise. Therefore, the contingency table for the Chi-square test in this case is similar to the one discussed above. The following table shows an example of the contingency table with combined variables $X_1 X_2 X_3$ and the target variable Z.

Table A.4: The 2×2 contingency table for the combined variable

	$Z = 1$	$Z = 0$	Total
$X_1 = 1, X_2 = 1, X_3 = 1$	n_{11}	n_{12}	n_{1*}
Others	n_{21}	n_{22}	n_{2*}
Total	n_{*1}	n_{*2}	n

A.1.3 Odds Ratio for Association Test

The CR-CS method introduced in Chap. 5 uses odds ratio to assess the association between variables.

With two binary variables X and Z and their contingency table (see Table A.1), the odds ratio of X and Z is defined as:

$$\omega = \frac{n_{11} n_{22}}{n_{12} n_{21}}$$

Given a threshold, e.g. normally 1.5 or 2.0, we can conclude that X and Z are associated if the odds ratio value is greater than the threshold, and vice versa. For example, given the contingency table in Table A.2, the odds ratio of Female and Low salary is 0.703. If we set the threshold as 1.5, the two variables Female and Low Salary are not associated (since $0.703 < 1.5$). They are not negatively associated either, since $0.703 > 1/1.5$

Alternatively, with a confidence level, we can find out its corresponding critical value z', then the confidence interval or the lower bounds and upper bounds of the odds ratio can be calculated as [1]:

$$\omega_- = \exp(\ln \omega - z'\sqrt{\frac{1}{n_{11}} + \frac{1}{n_{12}} + \frac{1}{n_{21}} + \frac{1}{n_{22}}})$$

and

$$\omega_+ = \exp(\ln \omega + z'\sqrt{\frac{1}{n_{11}} + \frac{1}{n_{12}} + \frac{1}{n_{21}} + \frac{1}{n_{22}}})$$

If $\omega_- > 1$, the odds ratio is significantly higher than 1, hence two variables are associated.

In the above example, when the confidence level is 95 %, the odds ratio is in the interval [0.9355, 2.164]. As ω_- is less than 1, we consider that the two variables Female and Low Salary are not associated.

A.1.4 Unconditional and Conditional Independence Tests Using R

In this section, we use the data set for the Bayesian network shown in Fig. 2.1, which can be downloaded from:

http://nugget.unisa.edu.au/Causalbook/

The data set contains eight predictive variables A, B, C, D, E, F, G, H and the target variable Z as described in Example 2.1. Assume that the data set has been stored in the *R* working directory. We will go through some examples of using Fisher's Z-transform for correlation and partial correlation from the *pcalg* package, and Chi-square tests from the *bnlearn* package. As shown below we firstly use the *condIndFisherZ* function in the *pcalg* package to test the conditional independence

```
> library(bnlearn)
> library(pcalg)
> data=read.csv("Example21.csv", header=TRUE, sep=",")
> data[1:5,]
  A B C D E F G H Z
1 1 1 1 0 0 1 1 1 1
2 1 1 1 0 0 1 0 1 1
3 1 0 1 1 0 1 1 1 1
4 1 0 0 0 0 0 0 1 0
5 1 0 1 0 0 1 1 1 1
#the first 5 rows of the data set

## condIndFisherZ <- function(x,y,S,C,n, cutoff)
## Value: Return TRUE if x is conditional independent
##        with y given S, and return FALSE otherwise.
## x,y,S: column indices of x, y, S
## C: Correlation matrix among variables
## n: number of samples
```

```
## cutoff: Cutoff value for a significance level.

C=cor(data)
n=nrow(data)
alpha=0.05 #significance level
cutoff=qnorm(1-alpha/2)

# Testing Ind(A,Z)
> condIndFisherZ(1, 9, NULL, C, n, cutoff)
[1] FALSE

# Testing Ind(A,Z|B)
> condIndFisherZ(1, 9, 2, C, n, cutoff)
[1] FALSE

# Testing Ind(A,Z|C)
> condIndFisherZ(1, 9, 3, C, n, cutoff)
[1] FALSE

# Testing Ind(A,Z|B,C)
> condIndFisherZ(1, 9, c(2,3), C, n, cutoff)
[1] TRUE
```

We can see from the results that A and Z are dependent conditioning on the empty set (unconditional), B, or on C. However, when conditioning on both $\{B,C\}$, A and Z are conditional independent. The results are consistent with the true causal structure in Fig. 2.1.

We now use Chi-square test from the *bnlearn* package.

```
## ci.test(x, y, z, test="x2", data)
## Value: return the Chi-square and p-value
## for the test of ind(x,y|z)

#bnlearn requires numeric or factor data types.
#We convert data in the data set to factors.
> for(i in 1:9){data[,i] = as.factor(data[,i]) }

#testing Ind(A,Z)
> ci.test("A", "Z", NULL, test="x2", data=data)

        Pearson's X^2

data: A ~ Z
x2 = 43.3043, df = 1, p-value = 4.685e-11
alternative hypothesis: true value is greater than 0

#testing Ind(A,Z|B)
```

```
> ci.test("A", "Z", "B", test="x2", data=data)

        Pearson's X^2

data:  A ~ Z | B
x2 = 9.8598, df = 2, p-value = 0.007227
alternative hypothesis: true value is greater than 0

#testing Ind(A,Z|C)
> ci.test("A", "Z", "C", test="x2", data=data)

        Pearson's X^2

data:  A ~ Z | C
x2 = 80.0481, df = 2, p-value < 2.2e-16
alternative hypothesis: true value is greater than 0

#testing Ind(A,Z|B,C)
> ci.test("A", "Z", c("B", "C"), test="x2", data=data)

        Pearson's X^2

data:  A ~ Z | B + C
x2 = 3.2395, df = 4, p-value = 0.5186
alternative hypothesis: true value is greater than 0
```

The tests return the Chi-square value (x2), the degree of freedom (df), and the p-value of the test. From the results, we can see that A and Z are dependent given the empty set, B, or C, as the corresponding p-values are small, i.e. less than 0.05. Meanwhile, the p-value for the test $Ind(A,Z|B,C)$ is large (0.5186), and therefore A and Z are independent conditioning on B and C.

References

1. J. L. Fleiss, B. Levin, and M. C. Paik. *Statistical Methods for Rates and Proportions*. Wiley, 3rd edition, 2003.
2. J. Schaefer, R. Opgen-Rhein, and K. Strimmer. corpcor: Efficient estimation of covariance and (partial) correlation. *R package version 1.5*, 7, 2010.

CPSIA information can be obtained at www.ICGtesting.com
Printed in the USA
LVOW07s1933270315

432329LV00002B/10/P

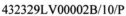